Geographical approaches to fluvial processes

"... of all places I have yet seen, I should prefer this in which to try an experiment I have long contemplated and have wished to have an opportunity to put into practice ".

      Robert Owen (1771 - 1858)

" Indeed, one could not choose a better place in which to study the process of waste, for one can examine the effects of rains, springs, and frosts, in loosening the sandstone by means of the hundreds of joints that traverse the face of the long cliffs, and can likewise follow in all their detail the results of the constant wear and tear of the brown river that keeps ever tumbling and foaming down the ravine ".

      A. Geikie, 1865

"... and these qualities are to be found in North Country built ships, such as are built for the coal trade, and in none other ".

      James Cook (1728 - 1779)

GEOGRAPHICAL APPROACHES TO FLUVIAL PROCESSES

edited by

Alistair F. Pitty

Contributors

Roger R. Arnett, Michael Bonell, Roger G. Cooper, John Crowther, Donald A. Gilmour, Lynne E. Frostick, Richard A. Halliwell, Anton C. Imeson, James A. Milne, Alistair F. Pitty, Ian Reid, J. Leslie Ternan, Roy W. Tomlinson, Dennis A. Wheeler, Philip A. Whittel, Andrew G. Williams, & Harry van Zon

Geo Abstracts
Norwich

© Alistair F. Pitty, 1979

ISBN   0 86094 026 8 paper
        0 86094 027 6 cloth

Published by Geo Abstracts Ltd.,
            University of East Anglia,
            Norwich, NR 4 7TJ, England.

CONTENTS

1	Introduction	1

PART I. Chemical weathering processes

2	Hydrological pathways and granite weathering on Dartmoor	5
3	Limestone solution on exposed rock outcrops in West Malaysia	31
4	Influence of contrasted rock types and geological structure on solutional processes in north-west Yorkshire	51

PART II. Hydrophysical processes

5	Runoff processes in tropical rainforests with special reference to a study in north-east Australia	73
6	Erosion processes in small forested catchments in Luxembourg	93
7	Relationships between plasticity, natural moisture conditions and surface stability of some slope soils near Helmsley, North Yorkshire	109
8	The use of differing scales to identify factors controlling denudation rates	127
9	Water levels in peatlands and some implications for runoff and erosional processes	149
10	Underground contributions to surface flow, as estimated by water temperature variability	163

PART III. Channel sediments and morphology

11	Drainage-net control of sedimentary parameters in sand-bed ephemeral streams	173
12	The analysis of pebbles on the bed of the Upper Wharfe, Yorkshire	203
13	The morphological relationships of bends in confined stream channels in upland Britain	215
14	The overall shape of longitudinal profiles of streams	241
15	Conclusions	261

To Professor H.R.Wilkinson

from some of his former students

EDITOR'S ACKNOWLEDGEMENTS

In addition to the contributors themselves, I would like to thank Penny, Edward and Alice Pitty for their forebearance during the compilation of this book. I am also indebted to Philip Larkin, Librarian of the University of Hull, for reading facilities, and to Mr. G.D.Weston, Assistant Librarian, for detailed information and advice. The staff of the Photographic Unit in Plymouth Polytechnic completed promptly the work on illustrations to several chapters. The introductory and concluding observations in this book, however, are solely those of the editor.

Alistair F. Pitty,
Cottingham,
East Yorkshire.
9th July 1979

CHAPTER 1 INTRODUCTION

Alistair Pitty

Number, weight, measure and experimental design are the cornerstones at the base of all pyramids of science. Little build-up towards an apex of scientific understanding can be achieved unless theories are framed, or practice re-aligned, by continual reference to the position and bearing strength of these foundations. Only recently has Geography begun to demand basic data, in numerical form, from its several specialist branches. The overall aim of this book, therefore, is to examine the main topics in fluvial geomorphology with reference to detailed counts, weighings and measurements from experimental designs which encompass real phenomena. Although the core of each chapter is original research data, three aspects of the presentation lend coherence to the work as a whole. First, the topics covered are arranged in a logical order, within three main parts, starting with the instantaneous and invisible process of chemical weathering in Part I. The mechanical transportation processes in Part II commonly involve visible movements of weathered materials, but they occur only after certain threshold levels are exceeded. Since the amount and manner of removal of weathered materials by fluvial processes depends largely on soil water and stream discharge, factors influencing volumes of soil water and runoff are emphasised. Part III focusses on the more permanent and generally larger-scale expressions of fluvial processes by illustrating and considering the sedimentary and landform characteristics of river channels. Secondly, a linking theme is developed within each of the main parts which emphasize the geomorphological purpose of the book. Thirdly, a distinctively geographical element is an integral part of each investigation.

THE GEOMORPHOLOGICAL PURPOSE

1.   Solutional processes

In Part I the solutional weathering of granite is the opening topic (Ternan & Williams, Chapter 2), partly because the research and discussion which have surrounded the origin of tors might epitomize much of the many ways of geomorphology in Britain since 1945. A prominent theme in such discussion, that of tropical weathering, is then taken up utilizing the advantages afforded by limestones for studies of natural chemical weathering (Crowther, Chapter 3). The advantages of relatively pure limestone include the simplification which its near-monomineralic composition ensures, together with the relative rapidity of reaction rates. A further advantage is the presence of bare limestone outcrops in the tropics, compared with the depths of regolith which commonly cover granites. Enlargement of the size of any study area beyond that of homogeneous outcrops,

Introduction

however, complicates the pattern in solutional processes. To illustrate this inevitable characteristic, Part I concludes with studies of the effect of impermeable and insoluble strata on an interstratified limestone (Halliwell, Chapter 4).

2. Hydrophysical and mechanical processes

Part II opens with the impact of a tropical downpour with the ensuing rates and routes of water movement on and beneath a rainforest floor being established (Gilmour & Bonell, Chapter 5). Some of the main ways in which litter and soil particles are dislodged in forested areas are then established (Imeson & van Zon, Chapter 6). Next, the critical degree to which rainwater, once incorporated into the soil, might induce mechanical instability in a soil mass and landsurface is examined (Cooper, Chapter 7). The net denudational effect of all mechanical processes combined is then illustrated, a study which inevitably summarises chemical contributions to denudation too (Arnett, Chapter 8). By this stage in the book it is clear that the points and volumes at which surface water actually begins to flow are not realistically described as streams of 'first order' as identified from blue lines on maps. Part II, therefore, concludes with exploratory reviews of two types of actual starting points to surface flow, as observed in peatlands (Tomlinson, Chapter 9) and at springs (Pitty, Chapter 10).

3. Channel sediment and morphology

The sequence of topics in Part III is clearly defined by two partially related criteria. The discussion moves from the smaller to the larger scales, and, concurrently, from readily modified phenomena to the increasingly durable features of the landscape. Initially, sedimentary characteristics of sands which may be changed with each flash flood are examined (Frostick & Reid, Chapter 11), followed by a study of the durability of larger sizes of particles in a pebble bedload (Whittel, Chapter 12). With processes and sediments considered, the individual studies conclude by examining two main morphological expressions of fluvial processes. The channel form and plan patterns which are created in gravel-covered floors of upland valleys are described (Milne, Chapter 13). Also the largest and most indelible expression of fluvial processes in the long term, the longitudinal profile of streams, is examined (Wheeler, Chapter 14).

THE GEOGRAPHICAL PURPOSE

The book examines fluvial processes ranging from the instantaneous raindrop splash to stream channel characteristics established over long periods of time. Despite this temporal range, only one aspect of Geomorphology is covered, a specialism which in itself is only one of many which can be incorporated into Geography. To ease the burden of breadth of study, it is increasingly vital that specifically geographical aspects of these many specialisms be identified. This book attempts to meet this urgent demand by exemplifying a range of distinctly geographical elements in its

enquiries and methods. For example, whilst significant
geological or temporal variations are stressed, 'variability'
is seen axiomatically to imply spatial variability. At
the broadest scale, the comparative method which
characterized the emergence of Geography as a separate
field of enquiry in the first half of the 19th century
persists. Contrasts and comparisons are drawn between the
temperate latitudes and the humid tropics since these two
major zones have continuous surface water as a conspicuous
and common feature. At a contrasted scale, actual worked
examples are broadly limited to sizes of area which are
readily comprehended if not encompassed by the individual's
gaze or walking range. This geographical scale is very
similar to that envisaged by many geographers in the 1920s
when they employed widely, if nebulously, the word
'landscape' to describe the sensuous object of their
studies. Critically it remains the scale which can involve
the contemporary student. It is the scale at which we can,
personally and literally, sense and grasp reality. It
follows that the results of investigations of actual land
surface phenomena are compared with natural occurrences in
other areas, rather than with expectations from theory or
with idealized predictions from models. Also, a comprehensive knowledge of the geography of places is used. Thus,
study areas are located where certain significant variables
are known or discovered to be clearly developed and where
other variables may be significantly absent. This distinctive approach applies at all scales from the major
zonal comparison to contrasts across a few metres of bare
rock or across a peat mire. Its degree of effectiveness
is evaluated in the conclusion (Chapter 15).

ANCILLARY THEMES

Emphasis on one aspect of a field of study does not imply
that other approaches are ignored or underestimated. This
is evident in three strong undercurrents to the main flow
of the book. First, to the geologist or to the historian,
time is usually the fundamental dimension. In contrast,
the three spatial dimensions are the geographer's main
concern. However, time remains as the inexorable fourth
dimension in which most geographical phenomena are to be
fully described and understood. Secondly, the stressing
of distinctively geographical aspects of scientific
investigations need not force the continuum of knowledge
into unrealistically compartmentalised components. In
fact, most of the following chapters develop themes in a
range of directions, reflecting the fields covered by the
contributors' qualifications. These include earth science,
ecology, environmental science, forestry, geology, hydrology
and soil science. In the context of this range of
experience, it is easier to identify where Geography
reaches its outer limits and to point to directions which
might lead back to its core. Finally, in exemplifying
a data-based methodology, no one technique or experimental
design is dogmatically employed. Indeed, a wide range of
considerations in devising or selecting appropriate
techniques is demonstrated. On the one hand, new equipment

Introduction

and sampling procedures have been readily adopted or adapted to tackle old, unyielding, problems. On the other, advances are made by a re-direction of the tried and trusted in exploring new fields.

CHAPTER 2 HYDROLOGICAL PATHWAYS AND GRANITE WEATHERING ON DARTMOOR

J. L. Ternan & A. G. Williams[*]

INTRODUCTION

Surface runoff may be an inevitable feature on bare-rock outcrops and of the ground surface in semi-arid or periglacial environments. The contrasted contribution to stream flow, originating as base flow emerging from a groundwater body, is seen particularly clearly in areas of well-jointed rock, like the Great Scar Limestone of northwest Yorkshire. A third main source of drainage water which may be significant is lateral, subsurface movement of soil water, favoured by impermeable horizons in the subsoil. The varying proportions of contributions to stream flow from these three main sources influences the water chemistry of streams.

The purpose of the present study is to measure the chemical characteristics of water at a spatial sequence of points along hydrological pathways in the Narrator Brook catchment, a small drainage basin some 4 km^2 in area situated on the western margin of Dartmoor (Fig 2.1). The approach adopted facilitates the evaluation of the principal controls on solutes in natural waters by identifying those points along the pathways where significant changes are observed and by relating such changes to the environmental characteristics which are well-developed at such points. First, it is commonly agreed that the earliest stage in a hydrological pathway is the atmosphere, and therefore the degree to which solutes are present in dry or wet atmospheric fallout is usually established. Secondly, water is considerably modified by passing through a vegetation canopy as throughfall and stemflow. Elements may be leached by rain from leaves and twigs (Tukey et al 1965) and salts may be trapped, with varying effectiveness, depending on canopy characteristics. These two sources must be strictly differentiated, if possible, as in the former case, elements are derived from within the catchment area as part of the biological cycling process, whereas the atmospheric salts are largely additions to the catchment's water chemistry, transported from a source beyond its perimeter. Thirdly, when water reaches the ground-surface, its pathways diverge. Some water will be returned to the atmosphere through evaporation or transpiration, although salts in the water will remain in the catchment or be removed along a different route. Some water will be involved in the nutrient flux and be released with transpiration or leaf decay. Although some of the remaining water may flow over the groundsurface as overland flow in some environments, most passes through the litter layer

---
[*] School of Environmental Sciences, Plymouth Polytechnic

Hydrological pathways and granite weathering

to join the soil water body. Biogeochemical processes within the litter layer and soil are important in bringing about chemical changes in the water. Decomposition of litter releases mineral nutrients, some of which may find their way eventually into streams (Bormann et al 1969). More significantly, litter decomposition releases carbon dioxide (Burges 1958) which may be dissolved in the infiltrating water to form carbonic acid which will aid dissolution of minerals within the soil and rock zones. Subsequently, some water which enters the soil zone may move vertically downwards to recharge the groundwater body. On slopes, however, with relatively impermeable subsoils, soil water may move laterally within the soil as interflow. During dry conditions one set of hydrological pathways may be utilized, but in wetter periods, other additional pathways may come into operation.

In the specific case of weathering processes on Dartmoor, many of the previous studies have relied on the examination of sediments and landforms, the direct or indirect physical expressions of weathering. In varying degrees, either past or present processes have been emphasized in the interpretation of the Dartmoor landscape and in explaining the origin of the decomposure of granite. The present chapter aims to contribute precisely to these discussions by establishing the rates of present-day decomposition processes and identifying the controlling factors. The approach is that of systematic chemical analysis of drainage waters from a carefully selected geographical range of sampling points, with particular attention paid to the mode and amount of loss of silicon (Si). This emphasis, apart from focussing on the key feature of the chemical decomposition of granites, also eliminates one source of variation which can complicate the study of ions which are released from the soil in small quantities only. Quantities of atmospherically derived silicon are negligible (Verstraten 1977). Therefore, the effects of variable entrapment of elements by vegetation can be discounted and it can be assumed that all silicon leaving a drainage basin is derived directly or indirectly from mineral weathering within the catchment.

THE REGIONAL SETTING

1. <u>Geology</u>

Geologically the Narrator Brook Catchment area is located on the granite of the Dartmoor pluton. These granitic rocks were emplaced in Upper Palaeozoic strata after the culmination of the Hercynian orogeny at the end of the Carboniferous period approximately 280 million years ago. It is believed they are part of a continuous ridge of granite extending from Dartmoor to the Isles of Scilly (Exley & Stone 1964) with the present outcrops merely the upward extensions of granite from this ridge (Fig 2.1). Many studies have indicated the composite nature of the Dartmoor granite, and four principal stages in the emplacement of the granite have been recognised (Brammall & Harwood 1923). Stage I is represented by relatively basic,

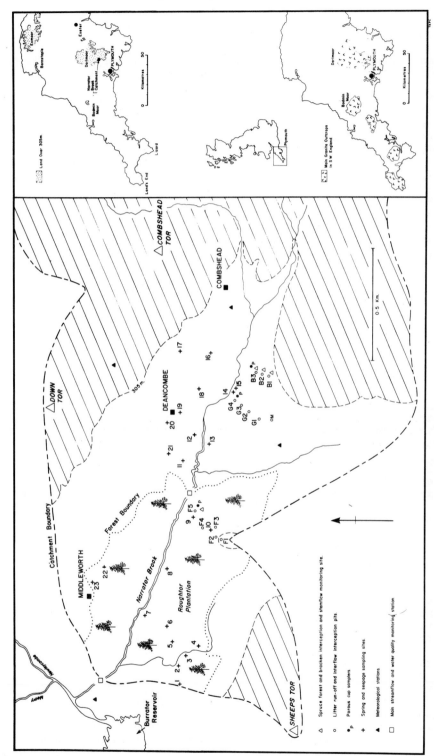

Figure 2.1 The Narrator Brook Catchment showing the investigation's location, sampling stations and hydro-meteorological network.

fine-grained biotite microgranite occurring as rounded inclusions in the two succeeding stages. Stage II, a coarse porphyritic biotite granite known as the 'Tor' or 'Giant' granite constitutes most of the plutonic rocks of south-west England. It is distinguishable by having potash feldspar phenocrysts averaging 3 to 5 cm in length in a coarse matrix of over 3 mm average grain diameter. Large crystals of quartz up to 1.5 cm across are also frequent and add to the porphyritic appearance of the rock.
Stage III, the 'Quarry' or 'Blue', granite generally underlies or is intruded into the Stage II granite. This is also a porphyritic biotite granite, although with a finer matrix of 2 to 3 mm average grain size and fewer feldspar phenocrysts attaining only about 2 cm length. Stage IV consists of dykes of finer-grained aplites which cross both Stage I and II granites and represent the last phase of intrusion.

The granite has been affected by several kinds of alteration. Evidence of tourmalinization is widespread with thin veins of secondary tourmaline occurring in joints. Kaolinization is extensive in the Lee Moor area of south-west Dartmoor and, from the depth of alteration of the granite as well as the occurrence of kaolinized rocks beneath an impermeable cover elsewhere in south-west England, is generally recognized as being of hydrothermal rather than weathering origin. Alteration of the granite has also occurred through weathering processes. These take place because the rock constituents are thermodynamically unstable in the presence of water and the biosphere. Hydrolysis by water containing carbon dioxide is the dominant process of clay formation and transformation in parts of the temperate zone as well as in the humid tropics (Brinkman 1977). Hydrolysis of the feldspars which are mainly converted to kaolinite is probably the most significant weathering process on Dartmoor. Frederickson (1951) has explained the process of hydrolysis as the penetration of a metal ion by hydrogen ions. As bonds become weakened, silicate groups pass into solution as silicic acid.

2. Relief

Evidence from denudation chronology studies suggests that the Dartmoor granite has been exposed to subaerial weathering and erosion since early Tertiary times (Brunsden et al 1964). Eastward-flowing rivers on an uplifted Cretaceous sea floor possibly removed the country rock and exposed the granite. This cycle of erosion culminated in the production of a postulated early Tertiary peneplain, remnants of which form the summit surface of Dartmoor at 580-518 m in the north and 503-457 m in the south. Studies of planation surface remnants and drainage networks have provided evidence of several erosion and weathering periods on Dartmoor during the Tertiary period. There is strong evidence of considerable periglacial activity on Dartmoor during the Pleistocene (Waters 1964). Evidence of frost action is widespread and includes the presence of blockfields and superficial head and solifluction deposits. This superficial material has in places been heaved and sorted by frost and moved downslope by solifluction processes. Patterned ground and other periglacial forms are common and occur at several places within the Narrator Valley including the slopes below Combeshead Tor and Sheepstor (Fig 2.1).

The Narrator Catchment ranges in elevation from 222 m OD at its outlet to 456 m at the highest point on the watershed, representing a total relief of 232 m. In its lower part the valley is markedly asymmetrical in cross-profile (Fig 2.2). The steeper slopes face north-north-east and although mostly between 10 and 20 degrees in angle some parts of this valley side attain higher angles. On the bracken slope profile, for example, maximum slope angles of between 25 and 30 degrees were recorded (Fig 2.3). These steep slopes are separated from low-angled (less than 5 degrees) moorland slopes by a marked break of slope at around 290 to 310 m OD (Fig 2.3). On the south-southwest facing valley side this break of slope is much less pronounced. Apart from the steep local slopes of the tors the lower valley side slopes rarely exceed 12 degrees and give way gradually to lower-angled moorland slopes at around 310 to 335 m OD. In the valley bottom the course of the Narrator Brook has been considerably disturbed by tin streaming and mining activity in previous centuries. The morphological effect of these processes is clearly demonstrated by abrupt breaks in many slope profiles, as seen, for example, in both the bracken and grassland slope profiles in Fig 2.3.

3. <u>Climate</u>

The prevailing climate is cool, wet and exposed to south-westerly maritime influences. Maritime air masses, especially from polar source regions, are experienced at all seasons and appear over the south-west of England for more than half the year. During the twelve months covered by this report (February 1977 to February 1978) the precipitation for the lower part of the Narrator Catchment was 1534 mm, a figure very close to the long-term average of 1585 mm from Redstone rain gauge 1 km to the south west. The wettest months recorded were October to January with rainfall in all four months in excess of 100 mm. The driest period occurred in late spring and early summer. Late August was unusually wet with almost 90 per cent of the monthly total occurring in the second half of the month, with a maximum intensity of 55 mm occurring in 18 hours on 24th August 1977. The hydrograph illustrated in Fig 2.8 resulted from this storm. Snowfall for the sampling year was slight, and in most years rarely lies for more than two weeks. Lowest mean daily temperatures occurred in January with a minimum of -2.15^0C recorded. Maximum temperatures occurred in July with a peak daily mean of 19.56^0C. The annual temperature range for the sampling period was 8.80^0C on available data for mean daily temperature. Potential evapotranspiration was approximately 650 mm.

Figure 2.2   View from Combshead Tor, looking south-west to the hillside on which the vegetation-slope transects were laid out. The bracken slope occupies the left of the view, the grassland much of the centre, and the south-east corner of the Roughtor Plantation appears to the right.

4.   Regolith

Like many drainage basins on Dartmoor, the granite of the Narrator Brook is overlain by a variable thickness of decomposed granite. During excavation of the trench for Burrator Dam, Burrator, solid granite was reached at depths of between half a metre and 12 m, with wedge-shaped fissures extending down to 33 m into the granite (Sandeman 1901). The decomposed granite was recorded as becoming more resistant with depth and merging into hard rock, in some cases abruptly. Within the Narrator Brook valley, the unconsolidated material at one point in the valley floor is known to be greater than 35 m thick, but no information is as yet available on depths of decomposed granite on the valley sides. This regolith frequently contains large, undecomposed boulders. Within the top few metres, layering is present. Such bedded deposits have been attributed by various workers to the action of downslope transportation processes, possibly accentuated under periglacial conditions (Waters 1964).

5. Soils

Two principal soil types are present within the Narrator catchment as well as being more widespread on Dartmoor (Clayden & Manley 1964; Harrod 1976). Acid brown earths of the Moretonhampstead Series are found on most sloping sites within the catchment, with typical profiles showing a fragipan at 70-90 cm depth and evidence of much biological activity. Infill chambers occur throughout the profile and one chamber 6-7 cm across was found at 80 cm depth. On higher flatter areas, podzolic soils (iron pan stagnopodzols) of the Hexworthy Series predominate. The principal minerals present in these soils include quartz, biotite, orthoclase, and plagioclase. The approximate clay mineralogical composition of one profile recorded by Clayden (1971) is vermiculite 45 per cent, mica 30 per cent, and kaolinite 25 per cent, although preliminary X-ray diffraction analysis of the clay fraction in the Narrator catchment indicates a higher percentage of kaolinite and only traces of vermiculite and chlorite.

6. Vegetation

The present vegetation is the result of many natural and man-induced changes which have occurred on Dartmoor in post-glacial times. Following an amelioration of climate about 7250 BC, cold tundra open heaths were largely replaced by woodland species (Simmons 1964). On many parts of Dartmoor, oak forest remained undisturbed for 4000 years, but deforestation began about 6000 BC either due to Mesolithic man or to natural processes, and blanket peat began to develop in areas of high ground cleared of forest. Subsequently, tin streaming in valley bottoms has destroyed much of the remaining damp valley woodland, grazing by animals has cleared more open land, and firing of heather moorland has initiated some peat erosion. Bracken spread is generally active and recently woodland areas have increased slightly due to afforestation, mainly with coniferous species. In the Narrator catchment, approximately 300 ha of woodland were planted in 1921. Sitka spruce (*Picea sitchensis*) is the principal species and the trees are now almost 20 m in height.

The vegetation of this part of Dartmoor shows a distinct gradation of community structure, dependent upon water and nutrient status (Kent *personal communication*). In the Narrator catchment, three slope transects were laid out as part of the experiment and provide a representative section through the principal communities present (Fig 2.3).

a) Grassland-moorland transect. Along this transect, the vegetation composition changes from acid moorland, dominated by *Molinia caerula* and *Agrostis setacea* on the

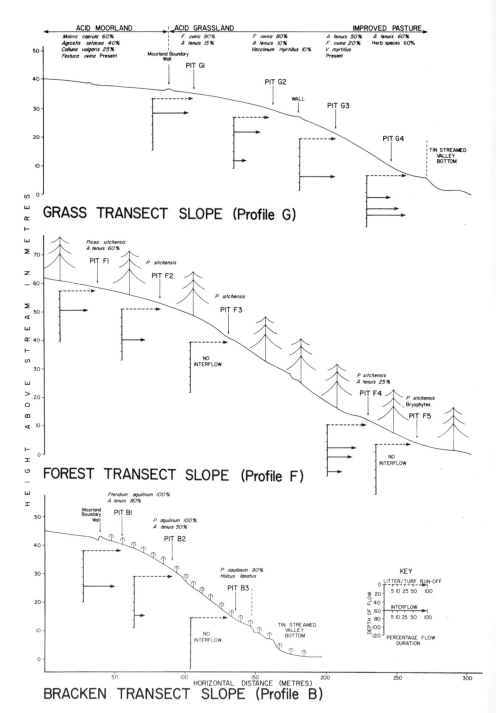

Figure 2.3 Three vegetation-slope profiles showing position of interflow and litter runoff interception pits. Depth and frequency of flow at each site are shown.

higher flatter parts of the slope to acid grassland downslope from the moorland boundary wall. This wall marks an abrupt change to an acid grassland community dominated by *Festuca ovina*. Downslope, a trend toward less acid species is demonstrated by an increasing dominance of *Agrostis tenuis* and associated herb species (Fig 2.3).

   b) Forest transect. Along the forest transect, the plant communities under the *Picea sitchensis* reflect the density of the tree canopy. Where the canopy is very dense, as at pits F2 and F3, only a litter layer is present. With a slightly less dense canopy, bryophytes are present, as at pit F5. With a more open canopy, *A. tenuis* comes in (F4 and F1).

   c) Bracken transect. On the bracken transect, invasion by *Pteridium aquilinum* appears to have occurred from the base of the slope where woodland grass (*Holcus lanatus*) is well-established under the bracken and a 50 per cent litter cover is present. Further upslope, *Agrostis tenuis* is increasingly dominant (Fig 2.3), and litter was not present at the top of the slope.

7.   Pedohydrology and geohydrology

   On the hillslopes, three modes of movement appear to occur. Firstly, in the acid brown earths, the fragipan horizon provides a significant barrier to water movement where present and a wet 'humose' layer occurs above the fragipan horizons. At lower horizons, the 'bedding' in the migratory layer of the regolith (Brunsden 1964) influences significantly present-day water movement on the valley sides. Secondly, on the flatter moorland areas, the iron pan in the profile is again particularly important in impeding vertical movement and in initiating both lateral flow of water within the soil profile and in supplying surface flows.

   Thirdly, as groundwater in undecomposed granite is guided by joints and fissures, the vein rocks in the granite may also act as controls on routes of groundwater flow. In a trench which was sunk during the building of the Sheepstor Dam, Burrator, the veins intersecting the trench were observed to act like deep land drains, throwing out a steady flow where they crossed the trench (Sandeman 1901). Numerous small springs and seepages occur within the Narrator Brook Catchment (Fig 2.1) and provide a major source of water supply to the stream. These springs generally occur at basal concavities on slopes, such as springs 15, 17 and 21. Several also emerge in valley floor locations, such as springs 11 and 18. Springs appear to emerge in distinct groups, with large areas with no springs or seepages between these groups. The distribution of springs may be explained in terms of discrete flows at different levels within the decomposed granite throwing out springs at both basal concavities and in gentler sloping valley bottom locations. Some springs issuing at one level begin almost immediately to lose water into their stream bed, such as springs 3 and 4, providing clear evidence of springs being fed by perched saturated zones.

Hydrological pathways and granite weathering

EXPERIMENTAL DESIGN AND METHODS
OF SAMPLE COLLECTION AND ANALYSIS

The experiment was designed to intercept and collect water samples for laboratory analysis at several points along the hydrological pathways. These points were ranged along three transects (Fig 2.3), each under a different vegetation cover. Transect G (grassland) was under acid grassland, dominated by *Agrostis tenuis* and *Festuca ovina* extending on to *Calluna* moorland. Transect B (bracken) was under acid grassland invaded by bracken (*Pteridium aquilinum*) and transect F (forest) was in a Sitka spruce *(Picea sitchensis)* forest (Fig 2.2).

1. Interception and throughfall and stemflow

Wet and dry fallout was collected using six 12.7 cm (5 inch) plastic funnels mounted 15 cm above the ground surface in polythene bottles, and positioned at intervals along one slope profile. Within the plantation, throughfall was collected in funnel rain gauges spaced 1 m apart (Fig 2.1). This network was designed to determine the effect of tree spacing and canopy density on solute concentrations in throughfall precipitation. Two further sites consisting of 8 gauges were also arranged at $90°$ compass points around selected trees, in order to obtain further data on solute inputs. Stemflow samples were obtained from collars fitted to trees at both ends of the throughfall gauge transects. Measurement of throughfall and stemflow under bracken was carried out using a method similar to that of Carlisle, Brown & White (1967). A 1 m x 1 m polythene sheet with side walls 15 cm high to prevent splash-in was secured to the ground before bracken growth commenced. In early summer, as the bracken emerged, slits were cut in the polythene to allow the growing shoot through. When the bracken was fully established, the junction between the bracken stems and the polythene was sealed. Although it was not possible to differentiate between stemflow and through fall water, this method proved satisfactory for the evaluation of water quantity and solute inputs.

2. Interception of soil water

A series of interception pits was established along the three transects, to collect water moving from the litter or turf layer and from the interflow, the soil water moving laterally downslope within the profile (Fig 2.4). These pits had polythene sheeting inserted at major textural horizons in the profile (Whipkey 1965) which channelled water into plastic guttering and then into polythene collecting bottles (Fig 2.5). The pits varied in depth from 1.3 m to 2.5 m, with the deeper pits being at the base of the slopes. Also, soil water samples were obtained from three pairs of cup soil water samplers (Parizek & Lane 1970; Wood 1973) installed at 90 cm (3 ft) and 182 cm (6 ft) depths, about 10 m to one side of the lowest pits of each transect. At all these water collection points, a bulk weekly sample for the period 14.2.77 to 6.2.78 was collected for analysis.

Figure 2.4   Example of soil pit, showing soil structure, horizons and the sampling assembly and recording equipment

3.   Springs

In the Narrator Brook Catchment, as on most of Dartmoor, much of the stream water is supplied by numerous permanent and intermittent springs and seepages. For the purposes of this study, a network of 23 sites was selected (Fig 2.1) and water samples collected at weekly intervals for the periods during which flow occurred. Although subject to similar controls to those directing interflow, these springs appear to be supplied from deeper levels. Air and soil temperature at the time of sampling are not considered to have had a significant effect on the water temperatures and inferences about the depths from which the springs emerge can be drawn from the standard deviation (sd)

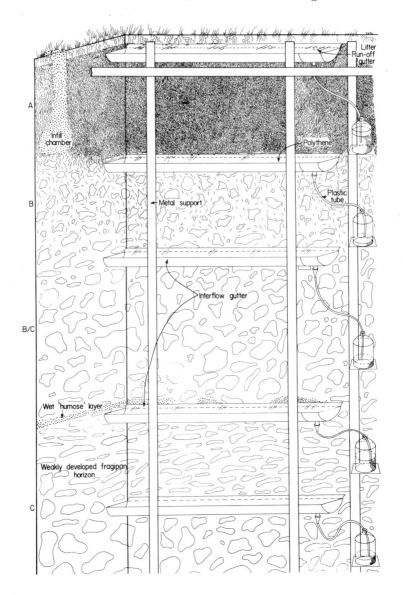

Figure 2.5 Pit assembly for the interception of litter/turf runoff and interflow.

of water temperature fluctuations, with a wide range suggesting near-surface flow whereas deeper sources are characterised by a narrow annual water temperature range. Thus springs from deeper levels would include springs 5(sd 0.12), spring 6 (sd 0.9), spring 12 (sd 0.13) and spring 13 (sd 0.26). In contrast, springs 3 and 11 (sd 1.37 and sd 0.88) would appear to be supplied by shallower sources. The more permanent springs tend to be located in valley bottoms whilst ephemeral springs are more commonly found on valley side positions. However, depth of flow, as suggested by narrow water temperature range, has little influence on frequency of flow. Some springs apparently supplied from deep sources dry up for long periods of time. Spring 20, with a temperature variation of sd 0.11 was dry for 16 weeks, and spring 21 (sd 0.34) was dry for 22 weeks. In contrast, spring 11, with a variation of sd 0.88 flowed for the whole year. This lack of association between depth of origin of flow and flow frequency is, however, the same pattern as was observed in the interflow pits, and similar controlling factors would appear to be in operation.

4. Stream channel and reservoir

Within the Narrator Brook stream channel itself, water samples were collected at 8-hour intervals using an automatic liquid sampler at the main stream gauging station at the lower end of the catchment (Fig 2.1). The final stage in the sequence is represented by the Burrator reservoir, and data discussed here is derived from monthly water analyses carried out by the South-West Water Authority.

5. Laboratory analysis

Water samples returned to the laboratory were normally analysed within a few hours of collection for pH and specific conductance. Silicon analysis was completed within 24 hours, using a Technicon II auto-analyser (Truesdale & Smith 1976).

HYDROLOGICAL CHARACTERISTICS OF INTERFLOW

1. Influence of soil profile discontinuities

Interflow occurs where there are well-developed textural breaks in the soil profile or regolith. The disposition of percolating water, on reaching a horizon of decreased permeability, can be illustrated by the case of pit G4. Here, measurable flow from just above the textural break at 92 cm occurred for 46 weeks. At a level 15 cm higher in the profile, flow was recorded for 27 weeks and at 40 cm above the impermeable horizon, flow was observed for only 14 weeks (Fig 2.3). This disposition of soil water is similar to the saturating upwards process observed by Weyman (1974). However, at some other sites, evidence suggested that, once a critical head of water has been built up above the first textural break, water begins to move through and downward until a deeper horizon of decreased permeability is encountered and lateral deflection of water flow and saturation upwards again occurs. Results from a pan inserted into the side of pit G4 at 2 m depth indicate that flow below the first major textural break occurred for only 14 weeks. At pits G2 and F4 this movement of water through to lower horizons in wet periods is demonstrated by the lower frequency of recorded flow at the lower horizons (Table 2.1 and Figure 2.3). It is likely that location on the slope transect is also important in this 'breaking through' to lower horizons, as sites lower downslope tend to have a larger catchment area for interflow. Whipkey (1965) also recorded seepage from shallower horizons occurring before that at deeper intervals, with the wetting front acting as an impediment to water movement.

2.   Influence of vegetation

The main influence of vegetation appears to be in the control on evapotranspiration and interception by the various plant covers which affects the amount of soil moisture present. Thus, along the grass transect, flow occurred for much of the year at all pits and interflow volumes were high, with amounts in excess of 5 l/hr per 1 m length in wet periods. On the bracken transect, interflow was much more variable, regular flow occurring from one pit only. Within the forest, interflow was also less frequent and more variable than under grass.

a) Grassland transect. The four interflow pits located along the grassland transect showed a marked tendency for the depth of the principal zone of interflow to increase downslope, as defined by frequency of flow occurrence (Table 2.1 and Figure 2.3). At G1, interflow was taking place from one horizon, just above an iron pan at 32 cm depth. Downslope, at G2, the principal flow zone was at 33 cm depth although, in the winter months of December and January, a lower wet weather route at 1 m depth also came into operation. At G3, flow was only recorded at 85 cm depth, although at G4 flow occurred at 92 cm and 77 cm. During very wet periods there was also flow at 51 cm.

b) Forest transect. Interflow occurrence was highly variable both with depth, as under the grassland, and also spatially (Fig 2.3). At F1, the principal flow zone was at 30 cm. At F2, no interflow was recorded for much of the year. When flow did occur, it was from 72 cm depth. At F3, no flow occurred at any level, nor at F5. There was flow at F4 at three levels, 109 cm, 76 cm and 53 cm.

c) Bracken transect. As beneath the forest, interflow occurrence along the bracken slope transect was highly variable (Fig 2.3). At B1, no iron pan was present, and flow was only occurring from the 85 cm horizon, a much deeper level than that recorded on the top grassland pit (G1) where an iron pan was present. No flow occurred at B1 for much of the period June 20th - October 24th, with the exception of a short period in late August-September, following a summer storm. At pit B2, flow only occurred on three occasions, at 90 cm depth, and after large storms. No interflow was recorded at B3.

VARIATIONS IN SILICON CONCENTRATIONS

The mean silicon value for stemflow/throughfall, litter layer, and interflow sites have been calculated, and the percentage of the sampling period when measurable quantities of water were obtained recorded (Table 2.1).

ACID GRASSLAND (TRANSECT G)

Precipitation: Trace of silicon

Site	G.1	G.2	G.3	G.4
Litter Run-off	0.38(94%)	0.47(77%)	0.77(80%)	1.68(88%)
Soil Interflow				
32 cm	0.54(70%)			
33 cm		0.71(41%)		
51 cm				1.02(34%)
77 cm				0.68(60%)
85 cm			0.92(78%)	
92 cm				0.79(84%)
100 cm		0.68(10%)		
Porous Cups				
91 cm	0.85(100%)			
182 cm	1.47(92%)			

SITKA SPRUCE (TRANSECT F)

Precipitation	no silicon detected	
Throughfall	trace of silicon	
Stemflow	0.24 mg/l	0.39 mg/l

Site	F.1	F.2	F.3	F.4	F.5
Litter Run-off	1.38(73%)	1.26(69%)	0.84(75%)	1.30(77%)	1.28(71%)
Soil Interflow					
30 cm	- (43%)				
53 cm				1.40(45%)	
72 cm		3.70(29%)			
76 cm				2.06(34%)	
109 cm				2.51(31%)	
Porous Cups					
91 cm	3.73(100%)				
182 cm	4.49(100%)				

BRACKEN (TRANSECT B)

Precipitation: no silicon detected

Site	B.1	B.2	B.3
Throughfall and Stemflow	0.59	0.40	0.50
Litter Run-off	1.80(89%)	2.84(83%)	1.64(77%)
Soil Interflow			
85 cm	1.07(59%)		
90 cm		0.80(5%)	
Porous Cups			
91 cm	1.63(94%)		
182 cm	1.99(94%)		

Table 2.1 Average silicon concentrations (mg/l), flow frequency and interflow depth for the three vegetation-slope transects. Flow frequency is indicated in brackets by percentage of sampling period flow was recorded.

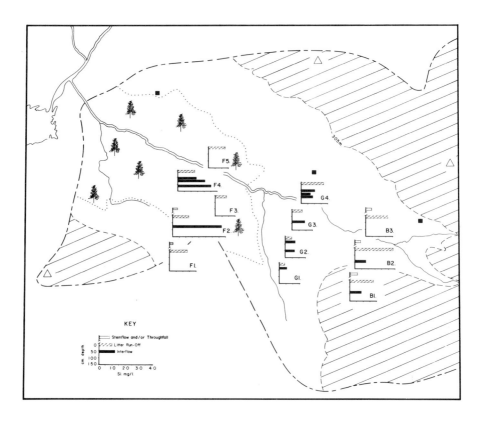

Figure 2.6 Silicon concentrations in stemflow, throughfall, litter runoff and interflow.

a) Grassland transect. In the acid grassland, the silicon concentrations in water increase along the pathway from lowest concentrations in water moving from the turf layer to increased concentrations in soil interflow, with highest concentrations in the soil water at the base of the slope, collected by the porous cups. Water collected from just below the turf layer contained increasing silicon concentrations downslope from 0.38 mg/l at G1 to 1.68 mg/l at G4 (Table 2.1, Figure 2.6). For all the points on this pathway the water contained lower silicon concentrations than at comparable points on the bracken or forest transects. In the interflow, silicon concentration also increased downslope. The mean concentration at G1 was 0.54 mg/l, increasing to 0.71 mg/l at the principal flow horizon at G2 and 0.92 mg/l at site G3. At the three depths at site G4, concentrations were 0.79 mg/l at 92 cm depth, 0.68 mg/l at 77 cm, and 1.02 mg/l at 51 cm depth. No obvious seasonal pattern in silicon concentration was apparent in water from the turf layer at site G1, G2 and G3, but G4 showed a distinct winter maximum, with a peak concentration at 4.5 mg/l in early January. Silicon concentrations in interflow from all pits also showed a seasonal trend with maximum concentrations at G2, G3 and G4 occurring in September and declining through the winter. At site G1, however, maximum silicon concentrations did not occur until January.

b) Forest transect. Silicon concentrations in water moving into and through Sitka spruce plantation increased throughout the pathway. Concentrations were lowest in stemflow and throughfall, increasing in the litter layer and interflow, before reaching highest concentrations in the soil water at the slope base in the porous cup samples (Table 2.1 and Fig 2.6). Silicon concentrations in throughfall were minimal and only samples obtained from throughfall gauges closest to the trees contained measurable concentrations. Water drainage through the needle litter layer showed much less spatial variability in silicon concentration than that recorded at the acid grassland sites, four out of five forest sites having had mean concentrations of between 1.26 and 1.38 mg/l (Table 2.1). Interflow, unlike that under acid grassland, showed little spatial pattern in silicon concentrations. Variations in silicon concentrations at F2 appeared to be dominated by flush-out effects after dry periods, with concentrations of 3.8 mg/l being recorded in late August after 15 dry weeks, and a concentration of 3.6 mg/l in late October after 7 dry weeks. At F4, concentrations were higher at deeper levels (Table 2.1 and Fig 2.6).

Seasonal variation in silicon concentrations was again present at some of the sites. Four of the forest litter sites, F4 being excepted, showed a fairly distinct seasonal pattern, with peak values occurring in late autumn and winter. No obvious seasonal trend occurred in the soil interflow waters.

c) Bracken transect. The combined throughfall and stemflow water obtained from the bracken plots contained silicon concentrations from zero to a maximum of 1.3 mg/l and at all three sites, the average concentrations (Table 2.1) were higher than those recorded under forest. Concentrations increased progressively through the growing season and remained high until the rapid decline of the bracken in late October. The litter water shows marked increases in silicon concentration over that found in stemflow and throughfall waters and at all three sites concentrations were higher than those recorded from the forest litter and most of the grassland litter sites (Table 2.1). In contrast with the two other transects, both interflow and porous cup samples contained lower silicon concentrations than the litter layer, although porous cup samples were again higher than interflow.

Silicon concentrations at all three bracken litter sites showed a pronounced seasonal pattern, with peak values of 9.2 mg/l and 9.3 mg/l occurring at B1 and B2 in January. By this time, the bracken canopy had gone, and these high peaks must be related to litter decay. Concentrations from February to April were again low, with grass species predominating during this period. Silicon concentrations in soil interflow at B1 followed a pattern of flush-out effects

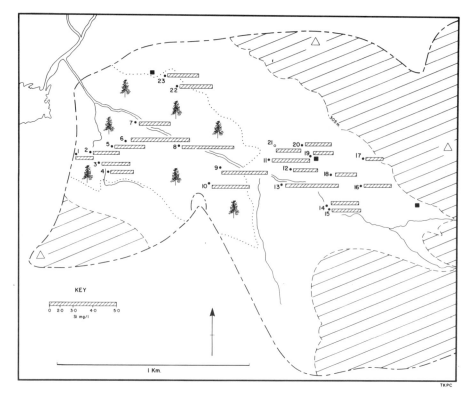

Figure 2.7 Silicon concentrations in spring waters

superimposed upon a broad seasonal variation. Although the mean concentration was only 1.07 mg/l, late August/early September storms resulted in a peak of 2.7 mg/l and a further peak of 4.2 mg/l was recorded on December 26th after five dry weeks.

   d) Springs. Average silicon concentrations in springs ranged from 2.55 mg/l at spring 17 to a maximum of 4.67 at spring 6 (Fig 2.7). No spring at any time contained a silicon concentration of less than 2 mg/l. Some seasonal pattern in the concentration may also occur. Of the eight perennial springs, those five having a higher mean silicon concentration in summer months appear to be moving at shallower depths, as indicated by water temperature variability (Table 2.2). The remaining three show a winter maximum in silicon concentration, and have a much lower recorded temperature variation (springs 5, 6 and 12) and are believed to be supplied from deeper levels.

   e) River and reservoir. In the main river, a distinct seasonal pattern of silicon concentration was observed. From February to July, concentrations were low, and on only one occasion did the concentration rise above 3.0 mg/l. From July to January, however, silicon concentrations were normally higher than 3.0 mg/l, with the peak of 3.6 mg/l occurring in late August/early September. Although the

Table 2.2   Summary of silicon concentration, water temperature and flow duration characteristics of sampled springs, based on all weeks in which flow occurred. Note: statistical analysis is based only on data from weeks in which flow at all springs, excluding flood springs, occurred.

Spring	Duration of flow	Mean Silicon mg/l	SD Silicon	Mean Water Temperature (°C)	SD Water Temperature
1	42%	2.61	0.31	8.70	-
2	58%	3.15	0.25	8.92	0.40
3	78%	3.24	0.35	7.98	1.37
4	100%	2.95	0.30	9.66	0.49
5	100%	3.35	0.31	9.21	0.12
6	100%	4.67	0.32	9.53	0.09
7	100%	3.38	0.40	9.37	0.65
8	100%	4.39	0.44	-	-
9	54%	4.14	0.27	-	-
**10	21%	3.75	0.08	-	-
11	100%	3.77	0.37	9.90	0.88
12	100%	3.04	0.26	9.83	0.13
13	100%	4.42	0.61	9.83	0.26
14	100%	3.20	0.28	9.66	0.59
15	81%	3.28	0.26	9.62	0.77
*16	100%	3.19	0.43	10.42	1.11
17	79%	2.55	0.33	9.48	0.67
18	81%	2.84	0.30	9.17	1.08
**19	21%	2.73	0.24	9.47	0.14
20	67%	3.10	0.32	9.56	0.11
21	54%	3.06	0.30	9.48	0.34
22	58%	3.49	0.36	9.46	0.39
23	58%	3.40	0.38	8.84	-

*Eight months samples only   **Flood springs

eight-hour sampling interval did not reveal any initial increase in silicon due to flushing with the onset of flood, dilution was always observed during flood events (Fig 2.8). The total run-off for the twelve month period of observation was 6,877,109 cumecs. Using the mean daily discharge and the mean daily silicon concentration, a total loss of silicon of 20,636 kg for the year is calculated. As denudation, this represents a silicon loss of 43.84 kg/ha/yr, equal to 93.84 SiO_2 kg/ha/yr. The artificial environment of the reservoir shows a seasonal low in March to early June, which may be related to diatom activity. This pattern may reflect a significant process in slowly moving water, as a close correlation between decrease silicon concentration and increase in diatom numbers from later March to May in the River Thames has been noted (Lack 1971).

SIGNIFICANCE OF VARIATIONS IN SILICON CONCENTRATIONS

a) The influence of vegetation. Silicon, although not an essential plant nutrient, is contained in varying amounts in plants, depending on species (Siever 1969). Plants are able to act as temporary reservoirs by accumulating silica, which is then redistributed with the eventual decay of the plant material. The microscopic opaline particles which are released can be detected in the soil (Acquaye & Tinsley 1965). Also silica may be exuded on to leaf surfaces or leached from the leaves by precipitation (Mina 1965). Possibly it is the higher transpiration rates of bracken which, leading to higher amounts of silica being exuded on to leaf surfaces, accounts for the high silicon concentrations in precipitation washing off bracken. Additionally, with the large leaf surface area, leaching of silicon from the bracken fronds by precipitation may be more effective than from the other plant species. Beneath grassland the progressive downslope increase in silicon concentrations in run-off from the turf layer coincides with an increasing dominance of *A. tenuis* (Fig 2.3) suggesting that different grass species may vary in their role as accumulators. Silicon concentrations in the shallowest levels of interflow at pit G4 also appear to be increased due to proximity to the plant cover.

In the spruce forest, the close contact between the stemflow and the rough-textured bark may explain why concentrations of silicon are higher in the stemflow than in the throughfall. Therefore, the differences in drainage water composition between the three vegetation covers may be related to variations in their effectiveness in accumulating, exuding and releasing silica.

In addition to biochemical processes, the vegetation cover is clearly a physical influence on the pedohydrological regime. The comparatively reduced evapotranspiration under grassland generates greater volumes of interflow, whilst the more extensive rooting of bracken and trees may open up channels through impermeable layers, which drain more effectively. Silicon concentrations appear to be closely related to such pedohydrological factors. Lowest overall concentrations of silicon occurred in interflow under the grassland transect and may be associated with greater frequency and amount of flow, with contact time with soil minerals correspondingly reduced. This may explain why in the porous pots at the base of the slopes, as well as in interflow, silicon concentrations were lowest under grass and highest under forest. Loughnan (1962) considered that the most important single factor controlling the breakdown of parent materials and the genesis of secondary products is the quantity of water leaching through the profile. Clearly, the type of vegetation cover is a significant

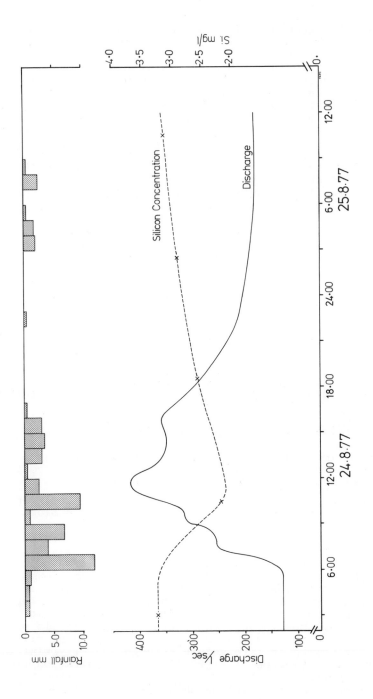

Figure 2.8 Silicon concentration and stream discharge during the flood of 24th and 25th August, 1977

control on such quantities and hence is an important, if indirect, control on variations in silicon concentrations.

b) Influence of soil profile depth and interflow occurrence. At some depth in a grassland soil profile, and within a given interflow zone, the decrease in silicon upwards from a textural boundary, as at pit G4, may be attributed to the more rapid transmission of wet-weather interflow moving above and more quickly than the more concentrated solution held up at the impermeable boundary. Where the wet weather route occurs at a lower level, as in pits G2 and F4 (Fig 2.3), the water contains high silicon concentrations owing to the less frequent flushing of these lower levels and to the presence of more concentrated water pushed through the textural boundary. Wet-weather interflow discharge may, therefore, include both shallow-moving, low-silicon concentration waters and more deeply flowing high-silicon concentration waters, or either.

c) Springs. Correlation analysis with mean silicon concentrations in water flowing from the springs as the dependent variable give a correlation coefficient of $\underline{r} = 0.67$ with a maximum slope angle leading to the spring (Fig 2.9). This variable explaining 45 per cent of the silicon variance may be interpreted as a reflection of the depth of water flow supplying the springs. Springs issuing from steeper slopes appear to be supplied from water from deeper, possibly bedrock, sources with a longer residence time and greater contact with the silicate minerals than that from springs emerging from gentle slopes and supplied by shallow moving water. All four springs with silicon concentrations greater than 4 mg/l (Table 2.2) issue from the base of slopes in excess of 13 degrees. This interpretation is supported by the water temperature data, springs from deeper sources and consequently with lower variation in water temperature emerging from steeper slopes (Fig 2.9). Hydrological factors influencing rate of flow, water residence time, and contact time between minerals and moving water would therefore appear to be the over-riding control on mean silicon concentrations of spring water.

Seasonal variation in silicon concentrations in spring waters may be related to both biological and hydrological factors. On shallow systems, peak silicon values occurred during the summer months, in contrast to deeper-seated springs where the peak did not occur until the winter. Although the possibility of high biological carbon dioxide production in summer may account for a summer maximum in solutional processes, the low bicarbonate concentrations measured, being less than 1 mg/l, suggest that hydrological controls predominate. It is therefore suggested that as wetting progresses through the winter, deeper levels only become flushed in late winter with a corresponding silicon maximum occurring in late autumn through to early winter. The interflow evidence of high silicon concentration flood routes supports this view. No evidence of dilution by rapid passage of water following heavy rainfall was detected at the springs.

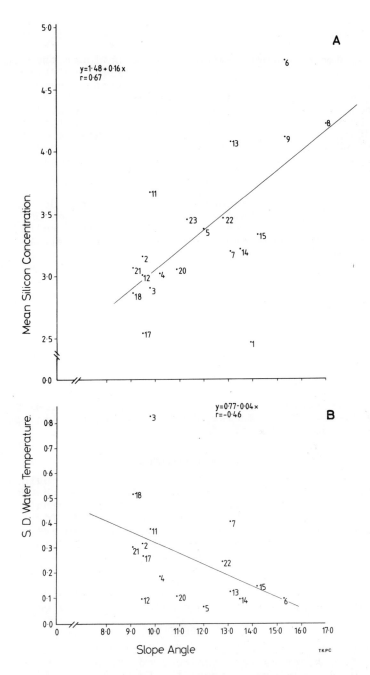

Figure 2.9 A. Relationship between local maximum slope angle and mean silicon concentration of springs
B. Relationship between local maximum slope angle and spring water temperature fluctuation, SD. (Standard deviation)

d) River. The pattern of silicon concentration in the Narrator Brook shows a broad seasonal variation, interrupted by short-lived dilution effects. In spring and early summer, concentrations tend to be lower than in the latter part of the year, a pattern similar to that recorded on many perennial springs. In contrast to the springs, however, marked dilution effects occur at all times during floods, as observed on 22nd August 1977 (Fig 2.8). This may be attributed to direct precipitation on the channel and to saturated soil drainage, particularly from the moorlands in the headwaters. Even in floods, however, silicon concentrations did not fall below 2 mg/l.

CONCLUSIONS

The results of the present investigation have demonstrated that contemporary chemical weathering processes on Dartmoor are active and continuous. Chemical weathering of the granite of this area is therefore not merely an historic process which is now largely inactive, as has been deduced by some geomorphologists studying the morphological evidence of landforms and sediments alone. The decomposition of silicate minerals is currently leading to the production of kaolinite where spring water is circulating more slowly at deeper levels and to gibbsite under very freely drained conditions nearer the ground surface. The presence of gibbsite is not, therefore, necessarily evidence of a former tropical climate. The present-day precipitation effectiveness and local drainage conditions are sufficient to weather and remove significant quantities of silicon, and can account also for the clay mineral characteristics of the insoluble residues. Except in flood periods, the dissolved load is greater than the mechanical transportation of suspended sediment.

ACKNOWLEDGEMENTS

We wish to acknowledge the cooperation of the South West Water Authority for allowing us to instrument the Narrator Valley for experimental work, and to thank Plymouth Polytechnic, School of Environmental Science, for its continued financial support for the project. We are also grateful to a large number of individuals, particularly Roger Cockerton, Trudy Crowder, Sarah Webber and Steve Johnson for technical, cartographic and photographic assistance, and the local South West Water Authority Manager, Mr. Gerry Taylor, for his cooperation on many occasions. Martin Kent helped with sections on vegetation.

REFERENCES

Acquaye, D. & Tinsley, J., 1965. Soluble silica in soils. In: *Experimental pedology,* ed. E. G. Hallsworth & D. V. Crawford (Butterwoth, London), 126-148

Bormann, F.H., Liken, G.E. & Eaton, J.S., 1969. Biotic regulation of particulate and solutional losses from a forest ecosystem. *Bioscience,* 19, 600-610

Brammall, A. & Harwood, H.F., 1932. The Dartmoor granites: their genetic relationships. *Quarterly Journal of the Geological Society of London*, 88, 171-237

Brinkman, R., 1977. Clay transformations: aspects of equilibrium and kinetics. In: *Soil chemistry, Volume VB. Physico-chemical models*. ed. G. H. Bolt

Brunsden, D., 1964. The origin of decomposed granite on Dartmoor. In: *Dartmoor Essays,* ed. I. G. Simmons (Devonshire Association), 97-116

Brunsden, D., Kidson, C., Orme, A.R. & Waters, R.S., 1964. Denudation chronology of parts of south western England. *Field Studies*, 2, 97-116

Burges, A., 1958. *Micro-organisms in the soil.* (Hutchinson, London)

Carlisle, A., Brown, A.H.F. & White, E.J., 1967. The nutrient content of tree stem flow and ground flora, litter and leachates in a sessile oak (*Quercus petraea*) woodland. *Journal of Ecology*, 55, 615-627

Clayden, B. & Manley, D.J.R., 1964. The soils of the Dartmoor granite. In: *Dartmoor Essays,* ed. I. G. Simmons (Devonshire Association), 117-140

Exley, C.S. & Stone, M., 1964. The granitic rocks of south-west England. In: *Present views of some aspects of the geology of Cornwall and Devon*, ed. K. F. G. Hosking & G. J. Shrimpton (Royal Geological Society of Cornwall, Penzance), 131-184

Frederickson, A.F., 1951. Mechanisms of weathering. *Bulletin of the Geological Society of America*, 62, 221-232

Harrod, T.R., 1976. Soils in Devon II: Sheet 5X65 (Ivybridge). *Soil Survey Record,* Harpenden

Lack, T.J., 1971. Quantitative studies on the phytoplankton of the Rivers Thames and Kennet at Reading. *Freshwater Biology*, 1, 213-224

Loughnan, F.C., 1962. Some considerations in the weathering of the silicate minerals. *Journal of Sedimentary Petrology*, 32, 284-290

Mina, V.N., 1965. Leaching of certain substances by precipitation from woody plants and its importance in the biological cycle. *Soviet Soil Science*, 6, 609-617

Parizek, R.R. & Lane, B.E., 1970. Soil water sampling using pan and deep pressure vacuum lysimeters. *Journal of Hydrology*, 11, 1-21

Sandeman, E., 1901. The Burrator works for the water supply of Portsmouth. *Proceedings of the Institution of Civil Engineers*, 146, 2-42

Siever, R., 1969. (Silicon) Abundance in natural waters.In: *Handbook of Geochemistry 2-1*, (Springer-Verlag, Berlin)

Simmons, I.G., 1964. An ecological history of Dartmoor. In: *Dartmoor Essays,* ed. I. G. Simmons (Devonshire Association), 191-215

Truesdale, V.W. & Smith, C.J., 1976. The automatic determination of silicate dissolved in natural fresh water by means of procedures involving the use of either α or β-molybdosilicic acid. *Analyst,* 101, 19-31

Tukey, H.B., Mecklenburg, R.A. & Morgan, J.V., 1965. A mechanism for the leaching of metabolites. In: *Radiation and istotopes in soil plant nutrition studies.* International Atomic Energy Agency, Vienna.

Verstraten, J.M., 1977. Chemical erosion in a forested watershed on the Oesling, Luxembourg. *Earth Surface Processes,* 2, 175-184

Waters, R.S., 1964. The Pleistocene legacy to the geomorphology of Dartmoor. In: *Dartmoor Essays,* ed. I. G. Simmons (Devonshire Association), 73-76

Weyman, D.R., 1974. Run-off processes, contributing area and streamflow in a small upland catchment. In: *Fluvial processes in instrumental watersheds.* ed. K. J. Gregory & D. E. Walling (Institute of British Geographers Special Publication No. 6), 33-43

Whipkey, R.Z., 1965. Subsurface stormflow from forested slopes. *International Association of Scientific Hydrology,* 10, 74-85

Wood, W.W., 1973. A technique using porous cups for water sampling at any depth in the unsaturated zone. *Water Resources Research,* 9, 486-488

CHAPTER 3 LIMESTONE SOLUTION ON EXPOSED ROCK OUTCROPS IN WEST MALAYSIA

J. Crowther[*]

INTRODUCTION

The chemical reaction between pure calcite, carbon dioxide and water is well-known under laboratory conditions (Garrels & Christ 1965; Picknett 1973; Picknett et al 1976; Plummer & Wigley 1976). In contrast, on a regional scale, geological factors explain much of the significant variations in the amount of solute in drainage waters. Nonetheless, if sufficiently small outcrops of limestone are located for study, many natural sources of variations are eliminated without introducing artificial controls or resorting to hypothetical assumptions. Exposed limestone outcrops, with minimal soil cover, provide the least complex of all solutional weathering environments. Solutional flutes developed on such rock surfaces form ideal sites for investigating the significance of a range of variables, since the characteristics of each micro-catchment can be determined unambiguously, with little chance of unknown factors being overlooked.

In the literature, there are few reports of the solute characteristics of waters draining bare rock outcrops. What little data exists is based on a small number or even a solitary sample and the possible significance of a spatial and temporal variation overlooked. In contrast, the present work is based on the study of run-off from 64 micro-catchments on limestone rock outcrops in West Malaysia, from which over 400 water samples were collected. The area selected for detailed investigation is Anak Bukit Takun ($3°18'N, 101°38'E$), a small limestone hill 18 km north of Kuala Lumpur (Fig 3.1). This hill was chosen because, although it is typical of the micro-relief of summit ridges over a broad area, it is relatively low and accessible, at an altitude of 170 m (Fig 3.2).

THE REGIONAL SETTING

Limestone outcrops are scattered throughout the northern half of West Malaysia. In most instances, particularly to the west of the Main Range, they occur as isolated tower karst hills, set amid broad alluvial plains (Crowther 1978).

---

[*] Department of Geography, St. David's University College, Lampeter

Limestone solution in West Malaysia

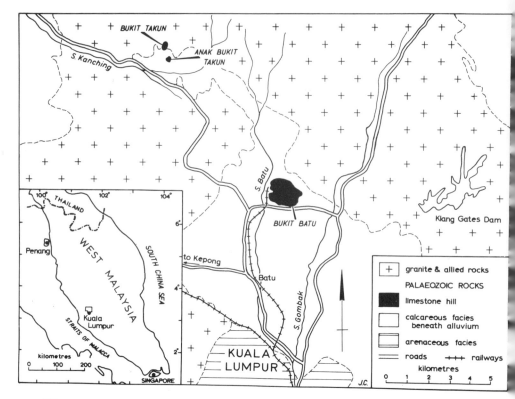

Figure 3.1   Location of the field area

1. <u>Geology</u>

The limestones are of Paleozoic age and were subjected to regional metamorphism and tectonic uplift during the Mesozoic. At this time the Malay Peninsula became part of a large cratonised block and, except for minor crustal deformations, the region has remained a stable landmass throughout the Tertiary and Cenozoic. The Kuala Lumpur limestones, of Silurian age, are up to 1830 m thick (Gobbett 1964) and are finely crystalline grey-cream marbles of low porosity and high mechanical strength (Jones 1973). The limestones in the study area are almost pure calcium carbonate, averaging 98.81 per cent $CaCO_3$, and since the measured range was 96.85 - 99.8 per cent, rock type may be regarded as constant.

2. <u>Relief and micro-relief</u>

Limestone is continuous beneath the adjacent alluvial plains and therefore the hills and towers of limestone are probably erosional remnants of formerly more extensive karst regions. The summits and ridge-tops, as exemplified by the Gunong Gajah-Tempurong massif (Crowther 1978), are invariably very dissected as a result of preferential solution along joint planes (Fig 3.3). The intervening

Figure 3.2 Anak Bukit Takun, the low hill in the middle distance, backed by the isolated tower of Bukit Takun

blocks of more massive rock are often pinnacle-like in form and may reach heights of over 20 m. The slopes below the ridges are boulder-strewn and often end in precipitous rock cliffs. Pronounced vertical solution and subsequent lateral undercutting at the base of the hills (Paton 1964), combined with the mechanical strength of the rock, are held to be largely responsible for present forms.

3. Climate

In equatorial regions, there is no marked seasonality of solar energy receipt and temperatures are uniformly high throughout the year. The mean annual temperature in Kuala Lumpur is $26.2°C$ (mean monthly range, $25.8 - 26.8°$). Aspect is another variable, significant in higher latitudes, which is essentially constant here, due to the high angle of the sun throughout the year. As a control on solutional processes, therefore, temperature in this area is constant, both temporally and spatially. The climate is of the equatorial monsoon type. Rainfall is fairly evenly distributed throughout the year, with precipitation maxima in April and October-November. Kepong, located 9 km south of Anak Bukit Takun, is the closest station for which long-term records are available and has a mean annual total over 33 years of 2440.8 mm. In lowland equatorial regions, much of the precipitation results from cells of convectional activity and throughout West Malaysia, storms of high intensity and short duration are characteristic, rather than long spells of continuous rain. Wind speeds are generally low and irregular in direction. In consequence, windward-leeward controls are absent and the fall of

Figure 3.3   Summit of Anak Bukit Takun

rain-drops deviates little from the vertical. This further reduces the effect of aspect as a factor in the solutional weathering of bare-rock outcrops.

When the investigation was carried out in July 1975, surface runoff from five storms was sampled. These downpours varied in magnitude from 11.4 mm in 170 minutes, 4.0 mm/hr, to 35.0 mm in 110 minutes, or 19.1 mm/hr. The maximum intensity observed was 21.8 mm in 45 minutes, a rate of 29.3 mm/hr.

4.   Soils and vegetation

Because of the high purity of the limestones and the locally steep topography mineral soils are thin or absent. Over much of the surface, the soil is restricted to patches of organic material which accumulate in poorly drained hollows and crevices on exposed rock surfaces. In the study area, soil depth was estimated at 1 m intervals along four 30 m-long transects which were spaced 10 m apart and the percentage of the surface covered by soil and litter was measured along each 1 m of section (Table 3.1). True mineral soils were absent, and five samples each of organic-rich mineral soil and decaying organic litter had pH values of 6.7, with a 6.1 - 7.2 range, and 5.6 with a 5.2 to 5.9 range, respectively.

Limestone solution in West Malaysia

	40 m x 30 m PLOT	64 SAMPLING SITES
Bare limestone)	66.3	24.1
Rock covered by lichen)		54.2
Rock covered by organic litter	7.9	7.4
Shallow soil, depth <10 cm	19.3	14.3
Deeper soil, depth > 10 cm	6.5	-
	100.0	100.0

Table 3.1 Percentage soil and litter cover on summit of Anak Bukit Takun

Vegetation on the hill-tops varies according to the soil cover. In more favourable areas, stunted and misshapen trees, typically *Vitex siamica, Memecyclon* spp. and *Garcinia* spp. (Henderson 1939), form an irregular and discontinuous canopy. A total of 50 trees (416/ha) with girth at chest height greater than 15 cm occurred within the 40 m x 30 m plot. The mean girth was 58.4 cm with the largest being 230 cm. Elsewhere, the bare rock surfaces support a wide variety of cremnophyte (rock crevice plants) and calcicole species, *Boea* spp, *Chirita* spp. and *Monophyllaea* spp. being common.

		pH	SPECIFIC CONDUCTANCE μmho.cm^{-1} 25°C	Ca^{2+} ppm	Mg^{2+} ppm	Na^+ ppm	K^+ ppm
RAINFALL	mean	5.46	10.0	1.12	0.14	0.17	0.13
	minimum	4.80	8.0	0.40	0.10	0.10	0.04
(10 samples)	maximum	5.80	13.0	2.40	0.20	0.20	0.19
THROUGHFALL	mean	6.15	17.9	2.46	0.30	0.20	0.65
	minimum	5.70	9.0	1.20	0.20	0.15	0.21
(20 samples)	maximum	6.60	33.0	4.00	0.50	0.30	1.80

Table 3.2 Chemical analysis of precipitation and throughfall

THE EXPERIMENTAL DESIGN

A major consideration in the experimental design within the 40 m x 30 m plot was the presence of a partial tree canopy, which altered the chemical quality of the rain water (Table 3.2) as well as intercepting and spatially redistributing the incoming precipitation. To ensure uniformity between micro-catchment areas, sites were selected only where there was no canopy cover. In addition, widely differing degrees of exposure to direct sunlight were thus avoided so variation in rock temperatures between sites was greatly reduced. Within these constraints, micro-catchments were then selected to encompass the complete spectrum of weathering environments observed in the study area. The

Limestone solution in West Malaysia

	MEAN	MINIMUM	MAXIMUM
Length (cm)	98.0	10.0	380.0
Width (cm)	21.9	7.0	84.0
Depth (cm)	10.8	2.0	32.5
Surface area (m^2)	0.29	0.01	2.52
Channel gradient (degrees)	49.4	23.0	85.0
% Organic litter/soil cover	21.7	0.0	95.0

Table 3.3 Characteristics of 64 sampling sites

main characteristics of the 64 solutional flutes studied are listed in Tables 3.1 and 3.3.

During storms, water was channelled into sampling bottles by means of transparent PVC tubing, attached to the rock surface by a small piece of plasticine. Both materials were tested to ensure that they did not affect the chemical quality of the water, and were installed using disposable polythene gloves, to prevent contamination of the water samples by salts from the hands. The sampling bottles used varied in size from 65 ml to 18 litres, depending on the micro-catchment size (Fig 3.4). Ideally, the sampling bottle should be of sufficient size to take the total volume of runoff, but at certain sites, particularly during storms of greater magnitude, the capacity of the sampling bottles was exceeded and water overflowed. In order to ensure that a representative sample was obtained under these circumstances, the tubing was inserted to one-third of the depth of the collecting bottle. This avoided bias in the sample towards water from the beginning or end of the storm, as would have occurred if the tube had been left in the neck of the bottle, or pushed to the bottom. Sampling bottles were installed in the field at the beginning of the experiment and samples were collected immediately after each storm. Solute concentrations were determined by standard methods of titration and spectrophotometry.

SPATIAL VARIATIONS IN SOLUTES BETWEEN MICRO-CATCHMENTS

Each micro-catchment received identical precipitation inputs and hence solute concentrations, averaged over the five storms for each site (Fig 3.5A), are comparable under precisely defined conditions. Specific conductance, which ranges from 30.3 to 173.4 μmho cm^{-1} at $25°C$, provides an estimate of the total electrolyte content. Calcium and magnesium are the dominant cations in the surface drainage water and this is reflected in the high correlation between total hardness and specific conductance ($r = 0.96$). In the study plot, calcium is the major cation, the ratio $Ca^{2+} : Ca^{2+} + Mg^{2+}$ being 0.96, the range being 0.88 to 0.99. The total hardness averages 51.8 ppm, with a range between 21.2 and 108.2 ppm. Total hardness, the sum of $Ca^{2+} + Mg^{2+}$ ions, expressed as parts per million $CaCO_3$, is used here as a measure of the amount of solution occurring within a particular catchment. The presence of non-alkaline hardness

Figure 3.4
Two typical bare-rock catchments with sampling bottles in place

is indicative of anions other than bicarbonate being present. These may be inorganic, such as chloride or sulphate, or of organic origin. Non-alkaline hardness varies between -3.1 and 22.3 ppm and averages 5.04 ppm. The negative values of non-alkaline hardness are all within the range of analytical error. Potassium, occurring in trace quantities in the bedrock, is one of the more important and mobile of nutrient elements in plants and is used here as a supplementary index related to the presence of substances of organic origin on the rock surface. The input of potassium from rainfall is 0.13 ppm whereas the net loss from the micro-catchments is 1.37 ppm, with losses ranging between 0 and 5.17 ppm.

1. Effect of soil/litter cover

The nature of the surface cover appears to be the fundamental factor controlling the solute content of surface run-off. On bare rock outcrops (Figs 3.4 and 3.5B), the average values of specific conductance, total, calcium and non-alkaline hardness and potassium are 61.10 μmho and 36.30, 34.89, -0.34 and 0.68 ppm respectively. Corresponding figures for sites with partial soil/litter cover (Figs 3.6 and 3.5C) are consistently higher, being 102.86 μmho and 63.17, 57.58, 8.97 and 2.10 ppm. No comparable data is available for bare rock outcrops in the humid tropics, but the average calcium hardness of run-off beneath a soil cover in Puerto Rico, based on five samples, was 60 ppm (Muxart *et al* 1969). In other climatic zones, however, the same authors report that calcium hardness of superficial runoff from bare limestone slopes varies from 37 ppm, with a range of 28 - 45 ppm, in Co. Clare, Ireland,

Limestone solution in West Malaysia

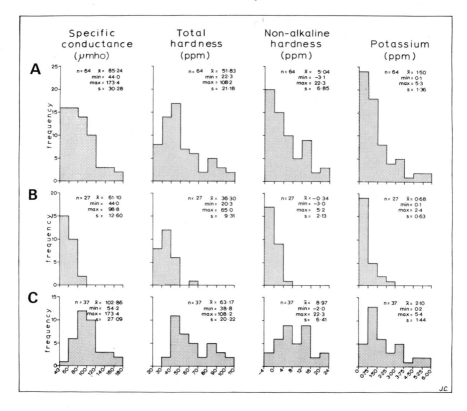

Figure 3.5   Solute content of superficial runoff averaged over 5 storms for each site
A. All sites (n = 64)
B. Bare rock catchments (n = 27)
C. Sites with soil/litter cover (n = 37)

to 71 ppm, with a range of 33 - 90 ppm, in Lapland, Norway. Twenty-eight samples of water from small seepages draining snowbanks on similar rock outcrops in Somerset Island, northern Canada, had a mean total hardness of 60 ppm.

Specific comparisons may be made between sites 4, 17 and 15 under fairly precisely defined conditions since all three micro-catchments are very similar in size and gradient but differ markedly in surface cover (Table 3.4A). Site 4 is bare of soil and litter, site 17 has a 5 per cent cover of leaf litter (pH 5.7), and site 15 has a 70 per cent cover of soil and litter (pH 6.5). The total dry weight in the last two catchments is 5.0 and 190 g, respectively. The total hardness, averaged over the five storms, is 28.6, 54.0 and 95.0 ppm for these three micro-catchments. Similar results have been reported from other limestone areas. Thus, in the Dachstein Alps of Austria at altitudes between 1400 to 2000 m, Bauer (1964) found that calcium hardness increased from 29 ppm on shallow runnels on bare rock outcrops to 49.4 ppm beneath a soil cover. Similarly, in the White Limestone area of Jamaica,

Figure 3.6
Micro-catchment with
a partial soil/
litter cover

Trudgill (1977) estimated erosional loss to be accelerated by more than five-fold beneath a thin 20 cm organic soil as compared with bedrock exposures.

On bare rock outcrops, the capacity of rainwater to dissolve limestone is primarily a function of the carbon dioxide content of the atmosphere and of temperature. The former may be assumed to be constant, at between 0.03 and 0.035 per cent, whilst the mean of 20 measurements established the water temperature in the study area at 29.5^0C. The consistently low non-alkaline hardness values suggest that chemical aggressiveness from sources other than carbon dioxide is slight. For bare-rock outcrops in general, 89 per cent of the sites have a total hardness in the range of 26 to 46 ppm. The single site which exceeds 49 ppm is also by far the largest catchment, with a length of 380 cm. With increased length of flow, there will be a corresponding increase in the average contact time between water and the rock surface and this will allow greater opportunity for the solutional potential of the rainwater to be expended. This factor may be considered as one possible explanation in interpreting the average values of the bare rock sites as a whole and the degree of variability observed between the individual catchments.

Where the rock surface has a partial or complete cover of soil or organic litter, the weathering environment is complex and exhibits much inter-site variability. Total hardness, with a mean of 63.17 ppm, ranges from 38.8 to 108.2 ppm. It is well-established that carbon dioxide levels in the soil and litter layers are higher than in the free atmosphere as a result of microbiological decomposition of organic matter, and that the increased chemical aggressiveness of rainwater is largely derived from this source (Drake & Wigley 1975; Rightmire & Hanshaw 1973). Organic acids released during decomposition may increase the

Site Categories	A. SOIL/LITTER COVER			B. CATCHMENT SIZE			C. CHANNEL GRADIENT		
Site number:	4	17	15	10	2	11	55	42	12
Morphometry:									
length (cm)	90	80	90	190	70	30	80	80	90
width (cm)	25	22	23	35	11	9	9	10	9
gradient (degrees)	53	50	55	60	64	64	46	54	66
Surface cover:									
% bare rock	87	90	30	80	60	75	40	30	40
% lichen	13	5	0	20	40	25	60	70	60
% soil/litter	0	5	70	0	0	0	0	0	0
Soil/litter, dry weight (g)	0	5	190	0	0	0	0	0	0
number of snails/100cm^2	0	0	0	0	0	0	3	5	3
Karst water (average of 5 storms):									
specific conductance (μmho)	50.6	89.3	141.2	58.4	50.2	59.0	73.2	72.0	56.6
total hardness (ppm)	28.6	54.0	95.0	36.7	29.9	33.9	37.6	44.6	32.2
non-alkaline hardness[1] (ppm)	0.4	8.8	14.2	<0.0	<0.0	<0.0	1.55	<0.0	<0.0
potassium (ppm)	0.6	1.6	1.5	0.4	0.1	0.4	1.0	2.4	0.4

[1] Values <0.0 are all within range of analytical error

Table 3.4   Details of sites used in examining the effects of surface cover, channel gradient and catchment size upon the properties of superficial runoff.

Table 3.5 List of the more close correlations (r) between total hardness and other variables

	ALL SITES (n=64)	BARE ROCK CATCHMENTS (n=27)	SITES WITH SOIL/LITTER COVER (n=37)
% soil/litter cover	0.622	--	
length			-0.282
surface area			-0.293
gradient	-0.418		
density of surface depressions		0.382	
density of snails		0.406	

solution rate, but their precise role is not clearly established. Their presence may increase solution, either directly through the formation of organic complexes with the metal cations (Roques 1969), or through reaction with the rock, or indirectly by means of the carbon dioxide released as they decompose (Bray 1972). In the study area, the non-alkaline hardness seems likely to come from organic sources, as an inorganic source is lacking. Of the average increase in total hardness of 26.9 ppm of the soil/litter covered micro-catchments compared with the bare rock outcrops, 17.6 ppm or 65.4 per cent is accounted for by an increase in alkaline hardness and therefore the source of the remaining one-third (9.3 ppm) may be attributed to organic acids.

The amount of soil/litter cover in the 64 catchments varies inversely with gradient (r = -0.69). In consequence there are significant correlations between catchment gradient and specific conductance (r = -0.46) total hardness (Table 3.5), non-alkaline hardness (r = -0.54) and potassium (r = -0.47). In order to examine the direct effect of channel gradient and other morphometric properties upon solute characteristics, discussion in subsequent sections is directed towards the 27 bare rock catchments.

2. Effect of micro-catchment size

It is uncertain, from theory, how rapidly solution occurs after contact between rainwater and the rock. There was no significant correlation between total hardness and catchment size in the 27 catchments. Three sites, 11, 2 and 10, were considered in more detail (Table 3.4B). They have similar gradients and proportions of surface cover but vary in catchment size (Table 3.4B), their lengths being 30, 70 and 190 cm respectively. No clear pattern emerges in the results. Total hardness increases from 33.9 ppm in the smallest catchment to 36.7 ppm in the largest, but is 29.9 ppm in the intermediate-sized micro-catchment. It seems that water flowing over only 30 cm of bare rock has sufficient time, on average, to attain a solute level which is barely exceeded in catchments of almost 2 m in length. Possibly,

Limestone solution in West Malaysia

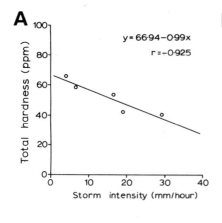
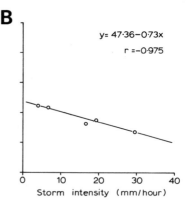
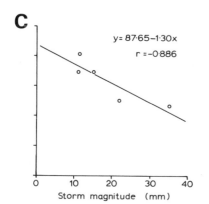

Figure 3.7
Temporal variations in solute content of superficial runoff - relationship between storm characteristics and total hardness, averaged for each of five storms.
A. All sites
B. Bare-rock catchments
C. Sites with soil/litter cover

depth of flow is a significant factor. The volume of flow increases cumulatively down the solution flute, and hence the depth of flow will vary directly with catchment size. Over the larger area, therefore, a smaller proportion of the run-off will be in direct contact with the rock face, and this tends to counter the tendency for solute content to increase with longer contact time with the rock.

3. Effect of gradient

In order to consider the possible effect of gradient, three bare rock micro-catchments, sites 12, 42 and 55, were selected for more detailed investigation. Their lengths, widths, and percentage cover are very similar, but their gradients range from 46 degrees to 66 degrees (Table 3.4C). Total hardness increases from 37.6 to 44.6 ppm as the gradient steepens from 46 to 54°. This proportional increase of 18.6 per cent is almost in the same proportion to the differences in rainfall receipt for each catchment. With channel lengths virtually the same, it follows that the horizontal distance through which rain falls is shorter the steeper the channel gradient. In the present case, the difference is 15.4 per cent. The higher hardness on the steeper gradient could therefore be attributed to the fact that a smaller amount of water falling on to steeper outcrops increases contact at the rock-water interface during a downpour compared with that in a thicker film of water draining off in which only a small proportion of the water molecules may come in contact with the rock. At even steeper angles, this effect may be counteracted by the reduced contact time associated with greater flow velocities. Thus, as the gradient increases from 54 to 66 degrees, total hardness falls to 32.2 ppm.

An implication of these tentative conclusions is that gradient is important as a control on the effectiveness of solutional processes at high angles, while at lower angles, volume of flow is the over-riding factor. The latter effect was also detected in temporal variations between storms (page 44) and changes during individual storms (page 46).

4. Influence of organisms

Locally, solutional processes on exposed rock surfaces may be influenced by the presence of lichen, moss or snails. In the present instance, snails appear to be particularly significant, and possibly are an important control. A high proportion of the bare-rock sites are pitted by circular depressions, 2-5 mm in diameter, 2-4 mm in depth, and in places exceeding 150 per 100 cm^2. Such forms occur even on vertical and overhanging rock surfaces and are, therefore, clearly not caused by rain-drop impact. There is a significant correlation ($\underline{r}$ = 0.83) between the density of depressions and of snails in the 27 catchments and both are related to total hardness (Table 3.5). The snails, of the species *Alycaeus*, occupy the cavities in the rock and in several micro-catchments the density exceeded 13 snails per 100 cm^2 of rock surface.

It is well known that snails produce acid secretions. Whilst the depressions have undoubtedly been caused by generations of snails, such a solutional process cannot, however, account directly for the enhanced solute content of the superficial runoff. Organic or inorganic acid secretions would increase non-alkaline hardness, but no such relationship was found. Most of the calcium and magnesium released from the bedrock is absorbed by the snails to form their shells and is only removed from the micro-catchment when the organisms die and their shells are washed away. The increased surface area and roughness resulting from unoccupied depressions provide a more likely explanation of the augmented total hardness values associated with snail activity. These would serve to increase the time during which water is in contact with the rock and produce greater irregularity and turbulence in the flow.

INTER-STORM VARIATIONS IN SOLUTES

Observation of temporal variations in the solute content of superficial runoff provides further insight into the solutional process. Solute concentrations were averaged over the 27 bare rock and 37 partially/totally covered

Limestone solution in West Malaysia

Table 3.6 List of the more close correlations (r) between rainfall characteristics and average solute content of superficial runoff for 5 individual storms (n = 5)

	A. BARE ROCK CATCHMENTS		B. SITES WITH SOIL/LITTER COVER	
	RAINFALL AMOUNT	RAINFALL INTENSITY	RAINFALL AMOUNT	RAINFALL INTENSITY
Specific conductance		-0.967	-0.834	
Total hardness		-0.975	-0.886	
Non-alkaline hardness			-0.992	
Potassium				-0.950

sites for each of the five storms, and their relationship with rainfall amount and intensity are given in Table 3.6.

In bare rock catchments, total hardness (Fig 3.7B) and specific conductance both exhibit a significant inverse relationship with rainfall intensity, but are unrelated to the amount of rainfall in a particular storm. Depth of flow and velocity of runoff are thus limiting since, with increased rainfall intensity, a smaller proportion of the runoff will actually be in contact with the rock surface, and the duration of contact will also be reduced. Under these conditions the total volume of water assumes no significance as atmospheric carbon dioxide, which provides the solutional potential, remains constant over time and is continuously available.

At sites with a soil/litter cover, potassium was additionally found to be inversely related to rainfall intensity. In contrast with the bare rock catchments, solutes also vary with the volume of precipitation input (Fig 3.7C). Specific conductance, total hardness, and, especially, alkaline hardness, thus decrease with increased storm magnitude. One possible explanation is that the carbon dioxide and more mobile organic constituents in the soil, which provide the potential for enhanced solution, become depleted during the storm as they are absorbed by percolating rainwater. The organic component, as reflected in non-alkaline hardness, would appear to be particularly influenced in this way (r = -0.99). Further evidence of this process may be observed in the temporal patterns of solutes during individual storms.

TEMPORAL CHANGES IN SOLUTES DURING DOWNPOUR

To investigate the temporal pattern of solute changes, water samples were taken from the nine catchments detailed in Table 3.4 at 10-minute intervals during a single storm. The observed chanced to have the highest intensity of the five storms studied, as 21.8 mm of rainfall fell in 45 minutes. The amount of precipitation over the five 10-minute periods (Fig 3.8) varied from 9.2 mm (55.2 mm/hour) in the second to 17.4 mm/hr in period 4.

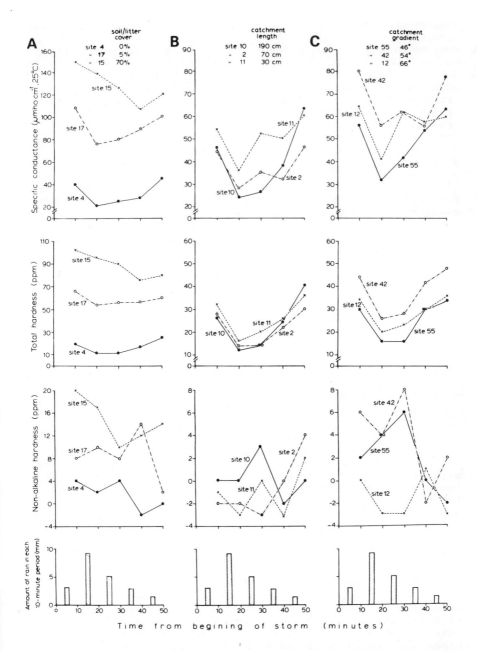

Figure 3.8 Temporal variations in solute content of superficial runoff - solute variations in selected catchments (detailed in Table 3.4) during a storm of 21.8 mm in 45 minutes.
  A. Selecting for soil/litter cover
  B. Selecting for catchment length
  C. Selecting for catchment gradient

Specific conductance and total hardness in all nine catchments are greatest at the beginning and the end of the storm. The high solute content at the end of the storm corresponds with the period of lowest rainfall intensity and may be largely attributed to a reduction in the rate and depth of flow. In the initial phase of the storm, some additional factor is operative since, in eight of the nine catchments, total hardness was greater in the first period than the fourth, when precipitation intensity was lower. Rainwater at the beginning of the storm may be taking up calcite materials deposited either as the tail end of the previous storm dried up, or developed between storms by solution and re-precipitation during alternating phases of condensation and evaporation. The latter possibility seems likely, since where rock surfaces are sheltered from direct raindrop impact or surface runoff, a white patina of re-precipitated calcite is ubiquitously developed.

1. Effect of soil/litter cover

As with inter-storm variations, the nature of the surface cover is the major factor controlling the temporal pattern of solutes during an individual downpour. On bare-rock outcrops (Fig 3.8B-C) and site 4 (Fig 3.8A), specific conductance and total hardness show clear correspondence with rainfall intensity. Site 4 is taken as typical of such catchments, with total hardness falling from an initial value of 20 ppm to 12 ppm during the period of maximum intensity, and rising again to 26 ppm at the end of the storm. Non-alkaline hardness, in common with the other six bare rock catchments, varies little from zero and exhibits no recognisable temporal pattern.

In contrast, at site 15, with a 70 per cent soil/litter cover, specific conductance and total hardness minima occur in the fourth period and non-alkaline hardness shows a distinct trend, falling regularly until period three and increasing at the end of the storm. These results conform with the pattern observed between individual storms (Page 44) and provide further evidence that, as the storm progresses, the reserves of carbon dioxide and organic acids are gradually depleted and the solutional potential reduced. The slight increase in the last period of the storm is attributed to greater contact between the rock and the dwindling water flow.

2. Effect of micro-catchment size

During the first 40 minutes of the downpour, total hardness was consistently lower at sites 2 and 10 than at site 11, the smallest of the three catchments (Fig 3.8B). This difference was most marked in the third period of the storm when total hardness at site 11 was 20 ppm compared with 14 ppm at sites 2 and 10. It may be tentatively suggested that, under conditions of high intensity of precipitation, thicker and faster flowing water films characteristic of the lower parts of the larger catchments are sufficient to prevent the full immediate potential of the rainwater from being realized. At lower intensities, during the final period, greater length of flow and reduced flow velocity appear to favour solution in the largest catchment.

3. Effect of gradient

The trends in total hardness at sites 55, 42 and 12 are almost parallel throughout the duration of the storm (Fig 3.8C). Total hardness is consistently highest at site 42, the catchment of intermediate gradient and this conforms with the spatial pattern of average values considered above (page 42). The marked inverse relationship between total hardness and rainfall intensity observed at these, and other bare rock sites, during this storm, lends further support to the tentative conclusion reached earlier that the influence of gradient upon solute concentration is mainly attributable to variations in volume of flow.

RATE OF SURFACE WEATHERING

Rainfall was carefully monitored at Anak Bukit Takun from the 1st September 1974 to 31st August 1975 to provide a record of the magnitude, duration and intensity of individual storms. The pattern of daily rainfall over the year corresponded closely with that of Kepong (r = 0.428), and the annual total of 2233.9 mm does not differ significantly from the mean annual total of the latter. Solution loss may be predicted using the regression lines of total hardness against storm intensity and magnitude, respectively, for bare rock catchments and sites with soil/litter cover (Fig 3.7). During the period under consideration, the estimated rate of surface lowering attributable to superficial runoff is 29.98 mm/1000 years in bare rock catchments and 50.22 mm/1000 years at sites with a partial cover of soil/litter. These rates will tend to be accelerated by snail activity and reduced by evaporation. The balance between these two opposed tendencies will depend mainly upon the nature of the surface cover. In bare rock catchments where snails frequently occur in large numbers and where rapid runoff reduces evaporation, actual surface lowering is likely to exceed the predicted value. Conversely, the presence of a soil/litter cover will diminish the rate of runoff and allow greater opportunity for evaporation loss, thus reducing actual rate of lowering to a level below that predicted.

If it is assumed that the mean annual precipitation at Kepong is representative of rainfall at Anak Bukit Takun and that the magnitude, intensity and frequency of storms recorded in 1974-75 are typical of the field area, then estimates of longer-term surface lowering may be made. Under the conditions which obtain on the summit of Anak Bukit Takun, the rate of surface lowering varies from 32.75 mm/ 1000 years on bare rock surfaces to 54.87 mm/1000 years where a soil/litter cover is present. It is interesting to note that these figures are of the same order of magnitude as on the limestone pavements of north-west Yorkshire. Sweeting (1966) estimates the rate of surface lowering in the latter area to be 41 mm/1000 years.

CONCLUSIONS

The Anak Bukit Takun experiment was established in such a way that several sources of variation in solution processes were naturally absent or excluded by the procedures adopted in site selection. Despite this, the solute content of the runoff from the 64 micro-catchments studied varied greatly. The total hardness for each site, averaged over the five storms, varied between 22.3 and 108.2 ppm. When intra- and inter-storm variations are included, the observed range is further extended, thus greatly facilitating the study of the relative importance of those controls which remained effective within the experimental design.

The results indicate that solute variations are largely dependent on the degree of surface cover, solution rates being increased significantly beneath soil or organic litter. The total hardness of runoff from bare rock sites averaged 36.3 ppm, compared with an average of 63.2 ppm for those with some soil or litter cover. In addition, particularly in bare rock catchments, the solute content varies in relation to the proportion of runoff in direct contact with the rock and to the length of contact time. Both are influenced in a complex manner by slope angle, catchment size and snail activity. During storms, the generally inverse relationships between total hardness and rainfall intensity in bare rock catchments and between total hardness and storm magnitude at sites with surface cover are readily understood. Solute content is consistently higher at the beginning of a storm, resulting from the removal of material weathered or deposited in the inter-storm period, and at the end when the effect of less, more slowly moving water was identified.

By sampling at a large number of points and under a range of conditions, the rate of weathering can be estimated with some confidence. Under the conditions which obtain on the summit of Anak Bukit Takun, the rate of surface lowering varies from 33 mm/1000 years on bare rock surfaces to 55 mm/1000 years under a soil/litter cover.

ACKNOWLEDGEMENTS

The work was undertaken whilst the author was in receipt of a Natural Environment Research Council Studentship. Chemical analyses were carried out at the University of Malaya, and I wish to thank Professor Zahara Hj. Mahmud of the Department of Geography, and Professor N. S. Haile of the Department of Geology, who made their laboratory facilities available to me during my time in West Malaysia.

REFERENCES

Bauer, F., 1964. Kalkabtragungsmessungen in den Osterreichischen. *Erdkunde,* 18, 95-112

Bray, L.G., 1972. Preliminary oxidation studies on some cave waters from S. Wales. *Transactions of the Cave Research Group of Great Britain,* 14, 59-66

Crowther, J., 1978. The Gunong Gajah-Tempurong massif, Perak, and its associated cave system, Gua Tempurong. *Malayan Nature Journal*, 32, 1-13

Crowther, J., 1978. Karst regions and caves of the Malay Peninsula, west of the Main Range. *Transactions of the British Cave Research Association*

Drake, W.D. & Wigley, T.M.L., 1975. The effect of climate on the chemistry of carbonate groundwater. *Water Resources Research*, 11, 958-962

Garrels, R.M. & Christ, C.L., 1965. *Minerals, solutions and equilibria.* (Harper & Row, New York)

Gobbett, D.J., 1964. The Lower Palaeozoic rocks of Kuala Lumpur, Malaysia. *Federation Museums Journal*, 9, 67-79

Henderson, M.R., 1939. The flora of the limestone hills of the Malay peninsula. *Journal of the Malaysian Branch of the Royal Asiatic Society*, 17, 13-87

Jones, C.R., 1973. Lower Palaeozoic. In: *Geology of the Malay Peninsula: West Malaysia and Singapore.* ed. D. J. Gobbett & C. S. Hutchison (John Wiley-Interscience, New York), 25-60

Muxart, R., Stchouzkoy, T. & Franck, J., 1969. Contribution a l'étude de la dissolution des calcaires par les eaux de ruissellement et les eaux stagnantes. In: *Problems of the karst denudation,* ed. O. Stelcl

Paton, J.R., 1964. The origin of the limestone hills of Malaya. *Journal of Tropical Geography*, 18, 138-147

Picknett, R.G., 1973. Saturated calcite solutions from 10^0 to $40^0 C$: a theoretical study evaluating the solubility product and other constants. *Transactions of the Cave Research Group of Great Britain*, 15, 67-80

Picknett, R.G., Bray, L.G. & Stenner, R.D., 1976. The chemistry of cave waters. In: *The science of speleology,* ed. T. D. Ford & C. H. D. Cullingford, (Academic Press, London), 213-266

Plummer, L.M. & Wigley, T.M.L., 1976. The dissolution of calcite in CO_2-saturated solutions at $25^0 C$ and 1 atmosphere total pressure. *Geochimica et Cosmochimica Acta*, 40, 191-202

Rightmire, C.T. & Hanshaw, B.B., 1973. Relationship between the carbon isotope composition of soil CO_2 and dissolved carbonate species in groundwater. *Water Resources Research*, 9, 958-967

Roques, H., 1969. A review of present-day problems in the physical chemistry of carbonates in solution. *Transactions of the Cave Research Group of Great Britain*, 11, 139-164

Sweeting, M.M., 1966. The weathering of limestone. With particular reference to the Carboniferous Limestones of Northern England. In: *Essays in geomorphology,* ed. G. H. Dury (Heinemann, London), 177-210

Limestone solution in West Malaysia

Trudgill, S.T., 1977. The role of soil cover in limestone weathering, cockpit country, Jamaica. *Proceedings of the 7th International Speleological Congress* (Sheffield), 401-404

CHAPTER 4 INFLUENCE OF CONTRASTED ROCK TYPES AND GEOLOGICAL STRUCTURE ON SOLUTIONAL PROCESSES IN NORTH WEST YORKSHIRE

R. A. Halliwell[*]

INTRODUCTION

In considering the solutional processes of the Ingleborough area of north-west Yorkshire, the Officers of the Geological Survey remarked that the area "... presents examples of this kind of underground erosion which are second to none in the Kingdom for number, extent, and interest ..." (Dakyns et al 1890). Indeed, the area has always aroused a keeness of general interest (Halliwell 1974). A key to much of the area's unique character is its geology (King 1960), and the present chapter examines the effects of geological controls, such as contrasted lithologies, jointing, and faulting on solutional processes within the limestone to the west and south of Ingleborough. In the context of fluvial processes, it is considered logical to describe the strata in the order in which they are crossed by water flow.

LITHOLOGY

1. Millstone Grit and Yoredales

The summit of Ingleborough is capped by flaggy sandstones and pebbly grit, but much of the remaining highland in the catchments is underlain by repetitions of a succession of Yoredale strata. These recur as cyclothems which, when fully developed, cause drainage to move off sandstone, over shale and then into limestone at the base of each cyclothem before re-appearing as surface drainage when the sandstone at the top of the stratigraphically lower cyclothem is encountered. Around Ingleborough, the surface exposure of Yoredale limestones is, in fact, relatively small compared with areas further north, and so surface drainage is not commonly interrupted. Continuing in downstream directions, at or near the outcrop of a shale bed with limestone, the *Girvanella* algal bed (Garwood & Goodyear 1924), lines of often vertical potholes lead drainage down into the depths of the underlying Great Scar Limestone. Thus, the main physical influence of the largely impermeable Yoredale strata is to encourage precipitation to gather in streams which then concentrate the erosive potential of water at certain points on the limestone. This concentration of flow encourages the development of cave sinks and underground systems which, in turn, increases the concentration of surface runoff. Thus, the process is self-accelerating.

[*] Academic Office, University of Hull

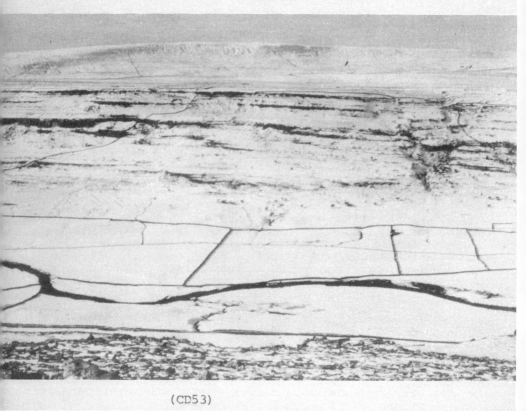

(CD53)

Figure 4.1 General view in north-west Yorkshire, looking north west from Ingleborough, showing typical relationsh between geology and fluvial processes. (Below the distant sandstone-capped ridge of Gragareth (627 m OD), surface drainage gathers on Yoredale strata before entering sinks (at about 380 m OD) in the uppermost horizons of the Great Scar Limestone. Swinsto Hole, for example (Fig 4.2E), is located close to the lowest point visible from that downsl wall which is second from the left-hand margin of view. The limestone lower slopes of Kingsdale its lies in 'dead ground' with Scales Moor dominating centre of the view. The limestone 'scars' show up in *black*. To the right, the effects of faulting o the line of Ullet Gill are clearly shown. Those risings included in the investigation by Halliwell (1977) are indicated by the position of the sampli site code as labelled on the margin of the view. T lower valley slopes, particularly below the line o the Old Road, are developed on the Ingletonian, wi the River Greta meandering across the superficial deposits on the valley floor of Chapel-le-Dale.

Solutional processes in N.W. Yorkshire

2. Great Scar Limestone

The Great Scar is a general name for a succession of limestones in north-west Yorkshire which vary from 100 to over 200 m thick in the Ingleborough area. The thin limestones of the upper beds form extensive pavements beyond the line of the main sinkholes. The underlying limestones are well-bedded, forming prominent scars (Fig 4.1) and reach a thickness of 104 m near Ingleton (Ramsbottom 1974). In underground exposures many bedding planes have been found to include shale horizons which are often of considerable lateral extent and with, in some cases, thicknesses of up to 2 m (Waltham 1971). The Great Scar Limestone itself consists mainly of fine-grained bioclastic limestones, usually of a pale grey to cream colour. The insoluble residue is commonly less than 2 per cent, with approximately half the matrix of foraminifera, shell and crinoid fragments, and the remainder either sparite or micrite. The relative proportion of sparite to micrite is significant (Sweeting & Sweeting 1969). The higher the percentage of sparite in relation to micrite, the lower the volume of available pore space. Thus solution may become restricted almost entirely to joint planes. Within the Great Scar Limestone, a geohydrologically important group of discontinuous, sparsely jointed and dense micrite horizons form the *Porcellaneous* band. At outcrop it may force water to the surface (Fig 4.2A). Where the lowest beds in the sequence filled up hollows in the pre-Carboniferous floor, shale bands are more prominent in the dark limestone, with a basal conglomerate including resistant pebbles of greywacke set in a calcareous cement.

3. Pre-Carboniferous rocks

As it may form a lower limit to water circulation within a mass of soluble rock, the strata beneath limestones are a significant consideration. In north-west Yorkshire, the limestone is underlain by the Ingletonian Series, a thick group of greenish, siliceous slates with intercalated resistant bands of silts and sandstone or greywackes. Rb-Sr data yield an age of 505 ± 7 m.y. for cleavage development, suggesting their deposition in the Cambrian or early Ordovician (O'Nions *et al* 1973). Isoclinal folding possibly occurred before 450 m.y. and uplift and erosion occurred by this time. In the 3.5 km exposure on the floor of Chapel-le-Dale (Fig 4.2A), the coarser beds cross the valley in broad, subdued ridges, showing a consistent NW-SE strike, whilst the finer beds are hidden (Dunham *et al* 1953). It is reasonable to conclude that much of the Carboniferous tract between Chapel-le-Dale and the other inliers to the east is underlain by the Ingletonian Series. In addition, it seems that a more extensive belt of magnetic rocks, probably metamorphic or extrusive rocks, underlies the Ingletonian rocks (Bott 1967). Also, a negative gravity anomaly which covers most of the Askrigg Block has now been proved by borehole test in a tributary valley of Wensleydale to be a granite batholith. Encountered at a depth of 510 m and yielding an Rb:Sr age of 400 m.y. (Dunham 1974), this intrusion of the Wensleydale granite is believed to explain

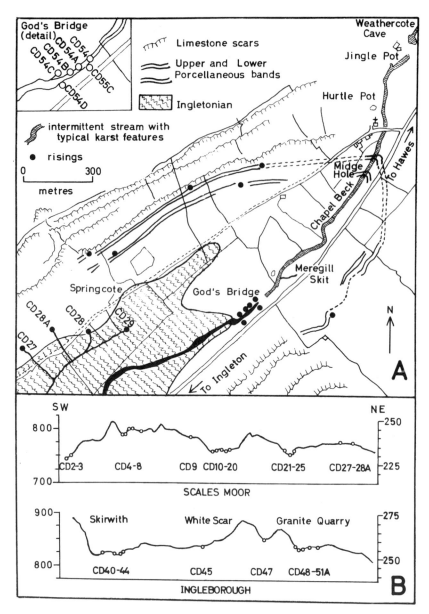

Figure 4.2 Lithological influences on underground drainage in Chapel-le-Dale and location of risings and typical karst features in the vicinity of Chapel-le-Dale. A. Influence of *Porcellaneous* bands on the location of minor risings and of the unconformity between the Great Scar Limestone and the underlying Ingletonian on risings and major resurgences. The reach of Chapel Beck between Weathercote and God's Bridge has many karst forms and geohydrological features typical of the classic underground rivers in the type-areas in Yugoslavian Karst. B. Influen of the 'fossil' relief of the Ingletonian outcrop o the location of risings on the north-west (Scales Moor) and south-east (Ingleborough) valley-sides of Chapel-le-Dale.

the rigidity of the pre-Carboniferous basement which has
left the overlying Carboniferous strata so conspicuously
unaffected by folding and faulting.

GEOLOGICAL FACTORS FAVOURING
VERTICAL MOVEMENTS OF GROUNDWATER

Marr introduced the concept of the area north of Craven
Faults as a rigid block which has shown a unity of
behaviour since Lower Paleozoic times and the massif of
the Askrigg Block forms part of the least-disturbed region
of the northern Pennines (Bott 1967).

1. Joints

In the southern part of the Askrigg Block, the dominant
direction of the jointing is parallel to the North Craven
Fault system, between N $50°$W and N $60°$ W (Wager 1931).
This direction coincides approximately with that of maximum
compression, deduced from the folding in surrounding areas.
North of Ingleborough summit, the direction swings round to
the trend typical for the northern Pennines, between
N $10°$ W and N $20°$ W. A second set of joints occurs at right
angles to the dominant set. It might be assumed that, since
laboratory shearing experiments induce angle of shear which
are much less, this highly distinctive rectangular jointing
pattern is anomalous. However, it has been proved that under
increased hydrostatic pressure, angles of shear do approach
$90°$. Thus, the distinctive, rectangular joint pattern of the
Great Scar Limestone was probably made when this rock was
deeply buried and so under great hydrostatic pressure,
possibly after the deposition of the whole Carboniferous
series (Wager 1931).

The influence of joints on the geohydrology of the
Ingleborough area struck members of the Yorkshire Geological
Society in their classic study of the water flows around
Ingleborough (Carter & Dwerryhouse 1905). They suggested
that the "... flow of the underground waters of the area
under consideration is radially outwards from the high ground,
but is profoundly affected by the direction of the joints in
the limestone, which in many cases considerably modify its
course ... So strong is the influence of the joints in the
limestone upon the direction of flow of underground waters
that there are several instances in which streams are carried
beneath a surface watershed so as to emerge in a different
drainage basin from that in which they took their rise ..."
Many cave systems, both in detail and at larger scales,
reveal the joint-control on underground passage development
(Fig 4.3A).

The distance apart and development of joints varies
considerably from place to place. An example of the coarser
extreme is found on the valley side above Weathercote, in
Chapel-le-Dale, where the joints are so widely spaced that
single blocks of limestone may be as much as $250m^2$ in surface
area. Lithology is significant, with the coarser-grained
limestones having a higher joint frequency than the fine-
grained carbonate mudstones (Doughty 1968). Cave systems
only reach to any depth where there is a reasonable density

of joints which pass through the shale horizons within the limestone succession (Waltham 1970). Also, the frequency of deep joints is related to the proximity of the Craven Fault zone for, with the exception of deep, fault-guided potholes, the depth-length ratios of the vadose cave system decreases steadily with increasing distance from the Fault. Thus, farther from the Fault, where deep joints are fewer, the caves demonstrate less vertical development. Wager (1931) considered that the modified direction of pressure, due to the dragging action along the North Craven tear-fault, produced a modification in the direction of the jointing in the neighbourhood of the Fault.

Strong joint control usually results in a rapid gain in depth, as seen in the majority of joint-controlled entrances to the Gaping Gill system. In the Alum Pot area, the shallow upper caves show only weak joint influences, partly because of the many joint directions in the area and the curvilinear joints near Borrins Moor. However, once the major NNW joints are reached, the passages lose height rapidly, forming Alum Pot proper. A similar relationship is observed further north in Washfold Pot. An exceptional case is the 1300 m long Dowber Gill passage in Wharfedale which, although aligned along one major joint and with a passage in places being up to 20 m high, has a low gradient.

The joints in the Basal Conglomerate are the same in direction as, and are sometimes even continuous with, those of the limestone above. However, in the uppermost 30 m of the Great Scar Limestone, which is less homogeneous and shows marked bedding, the jointing is less regular than in the massive limestone below (Wager 1931). Above, in the Yoredale limestones, jointing directions are very variable. Nonetheless, Yoredale limestones may contain pronounced joint-controlled caves, such as the Devis Hole Mine Cave (Fig 4.3B) and Windegg Mine Caverns in Swaledale (Ryder 1975). It is thought that very slow, retarded water flow through the limestone and along every joint produces a high-density graticule of passages (Palmer 1975).

2. Faults

The Lower Carboniferous area of Ingleborough is bounded to the south-west by major faults with a downthrow to the south-west. These Craven Faults have had a long and complex history, originating with Hercynian compressive forces, and are still active on a small scale. The greater part of the throw of the Outer Craven Fault and the formation of its continuation east-south-east as the South Craven Fault are regarded as the effect of the post-Permian earth-movements, with renewed sinking of the south Craven area in relation to the stable Askrigg block. Between Ingleton and Skirwith, the innermost member of the fault system brings Lower Palaeozoic rocks in contact with the Kingsdale Limestones. The latter beds are then brought against Coal Measures to the south-west by the outer fault, immediately to the north-east of Ingleton. By bringing the limestone against impermeable strata to the south, the Craven Fault zone acts as a barrier to water flow in that direction. This explains why there are several minor risings along the fault lines.

Solutional processes in N.W. Yorkshire

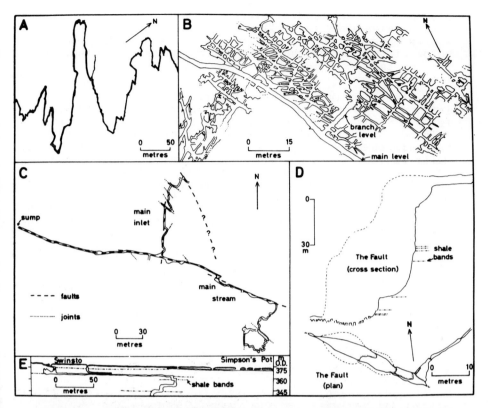

Figure 4.3   Structural influences on underground solutional activity. A. Larger and small-scale control of jointing on the development of passages in Smeltmill Beck Cave, Brough. B. Naturally enlarged joint network in the Central Maze area of the Devis Hole Mine Cave, Grinton, Swaledale. C. Fault and joint control on the plan of Tatham Wife Hole. D. Fault control on the cross-section and plan of Growling Hole, East Kingsdale. E. Control of regional dip on the development of some of the inlets to the West Kingsdale System (compare with limestone scars of Scales Moor, Fig 4.1). (These extracts from original maps have been reproduced following discussions with the principal surveyors involved. The originals are to be found in the 'underground literature' as follows: A. Yorkshire Underground Research Team, Report No.2 (1970), opposite page 42. B. Molywarps Speleological Group Journal No. 7 (1974), opposite pages 4 and 6. C. Dr. A. C. Waltham *in* London University Caving Clubs Journal, No. 6 (1968), page 20. D. Lancaster University Speleological Society Journal, Vol. 3 (1973), page 32. E. "West Kingsdale System Survey", published by the University of Leeds Speleological Association (1967)).

but there are no major risings, nor are there known caves running along the fault lines. However, caves like Storrs Cave have developed between the faults and those below Storrs Common show evidence of severe flooding and they may act as a major drain for the area between the faults.

Elsewhere in north-west Yorkshire, faults have influenced the pattern of several caves. On Scales Moor, in both Spectacle Pot and Growling Hole, there are chambers aligned along the calcited fault planes. In Growling Hole, the main chamber is up to 20 m long and 60 m high, (Fig 4.3D), whilst in Spectacle Pot, the main chamber is approximately 30 m long and 60 m high (Gascoyne 1973). These chambers contrast greatly with the entrance series to the caves, not only in size but also in direction. Both caves exhibit little vertical development until the fault plane is reached, where both lose height rapidly, dropping almost through the full thickness of the main limestone. In Tatham Wife Hole the majority of the cave passages is aligned along two intersecting faults (Fig 4.3C), but here the composition of the fault infills is a significant factor. The major fault, with a calcited infill, has been easily eroded to form a relatively spaceous cave along its length. The minor dolomite-infilled fault, on the other hand, whilst still being eroded to form a cave passage, has produced passages of very limited height and cross-sectional area.

Other examples of fault control include the long relatively straight large passage of Birks Fell Cave with its extensive collapse features. In the Gaping Gill system, a fault zone can be traced through the large Sand Caverns - Stream Chamber passages and along SE Passage (Glover 1974). The main chamber of Gaping Gill, at 145 m by 25 m by 35 m high being the largest known underground chamber in Britain, has been shown to be on the axis of a rotational or wrench fault, with a displacement of a *Porcellaneous* Band ranging from zero at the western end of the chamber to a 4 m downthrow to the south, at the eastern end. Many of the cave systems to the east of Gaping Gill, Juniper Gulf, Rift Pot, Nick Pot, descend rapidly down fault planes to almost the same altitude as the risings. The degree of fault brecciation is important, with Rift Pot consisting of a series of large loose chambers whilst in the less brecciated Juniper Gulf, the passages between the shafts are smaller.

Even where there are no known cave passages, fault lines appear to control the subsurface hydrology of the area. The large fault complex at Ullet Gill is probably the main reason why large volumes of flood water issue from these risings after wet weather. Similarly, in Dry Gill cave, although no accurate survey has been published, it is extremely likely that the main upstream sump is aligned along the fault line just to the south end of the known cave.

FACTORS FAVOURING LATERAL MOVEMENT OF WATER

1. **Dip**

Although the Askrigg Block has resisted folding in post-Carboniferous times, a regional tilt towards the north-north-east prevails. The influence of dip on underground drainage increases as the inclination of dip steepens. In the Ingleborough area the most significant feature is that, contrary to the NNE direction of the 4^0 regional dip, the surface drainage is to the south-west and south. Thus many predominantly vadose cave passages run generally down dip to the north until they become totally flooded at an altitude close to that of the risings. In this phreatic zone, cave orientation changes and the surface water flows southwards from the risings. The Meregill system is a classic example of this effect whilst the shallower Calf Holes-Browgill system is simpler, being basically simple downdip flow from sink to rising with some slight joint influences. As the dip increases, so its influence becomes more apparent and the extreme case of vertical dip may be seen in Barbondale where, as a result of their proximity to the Dent Fault, the strata are near-vertical. Although this structure has little control over the direction of water flow, the effect is the reverse of the usual Yorkshire cave's geohydrology, with Grid Pot containing a passage down a vertical shale bed, followed by a near-horizontal phreatic-tube passage along a joint. The influence of intermediate dips is best seen in Britain in the Mendip area where the long profiles of underground streams reveal a series of down-dip and up-joint loops.

In some areas of north-west Yorkshire, local folding is sufficiently strong to override the effects of the regional dip, and in these areas the major cave streams flow laterally down-dip, and then follow the plunge of the axes. This type of structural control has been described for the Manchester Hole - Goyden Pot - Nidds Heads system and many of the other caves at the head of Nidderdale and, in South Wales, on the development of Dan-yr-Ogof (Coase 1977).

2. **Superficial deposits and soils**

Precipitation falling on to a bare limestone pavement percolates vertically down through the rock. In many parts of north-west Yorkshire, particularly at higher altitudes, the limestone is covered with superficial materials, probably of solifluvial origin, overlain by peat or an organic-rich soil. Peat or subsurface soils with a high organic matter content readily absorb precipitation, but infiltration may then be held up by more clayey subsoils. This is the case in much of West Kingsdale where soil moisture measurements collected over Simpson's Cave and Swinsto Hole (Fig 4.3E) showed little evidence of accelerated soil drainage into the caves, even where the roof was only 2 m below the surface. The observations (Table 4.1), the high values being attributable to absorptive properties of peat, suggest that the clay horizon in the subsoil slows vertical drainage and encourages sideways waterflow to the small shakeholes which penetrate the soil cover down to

Solutional processes in N.W. Yorkshire

Table 4.1 Mean soil moisture percentage

Location	7.6	6.1	4.6	3.1	1.5	CAVE	1.5	3.1	4.6	6.1	7.6 m from cave
Swinsto[1]				552		469		596			
Swinsto[2]	491	564	524	468	497	478	438	483	520	492	466
Simpson's[3]	423	221	314	266	248	294	263	285	243	238	250

	SHAKEHOLE	5	8	11 m from shakehole
shakehole[4]	rock	318	460	543
shakehole[5]	rock	295	454	512

Sampling dates: 1 = 6.6.72   2 = 15.7.72   3 = 25.11.72
                4 = 13.7.72   5 = 15.10.72

bedrock. These act as drains in the same way as cave sinks, and their development could be self-accelerating.

The presence or absence of boulder clay can also directly affect subsurface hydrology by blocking passages. An example of a now-active stream sink which was, until relatively recently covered by boulder clay, is the Rat Hole Inlet to Gaping Gill (Rule 1910). It is a measure of the effectiveness of a boulder clay cover in favouring lateral movement of drainage water that the underground drainage from Alum Pot can cross beneath the river Ribble before resurging at Turn Dub. Such circumstances complicate the application of watertable concepts to the limestones of this area.

3. Shale beds in the limestone succession

After Simpson (1935) reversed the ideas of Tiddeman (1890) by stressing the significance of shale beds in arresting the downward movement of water within the limestone mass, many of the cave descriptions which followed noted the important part played by shale beds (Grainger 1938; Atkinson 1949). Myers (1948) expanded the concept when he observed that, in many cases, a passage started as a low opening on a thin shale bed and then cut down into the limestone below. Subsequently, Waltham (1970) has examined how main passage enlargement takes place in the limestones immediately below the shale in the vadose environment.

4. Lithological variations within the limestone

Due to the relative scarcity of joints cutting the micrite horizons of the Porcellaneous band, water flow tends to move horizontally along this band (Glover 1976). In Chapel-le-Dale, at outcrop, it may force water to the surface (Fig 4.2A). In the Gaping Gill area, the Porcellaneous band has been shown to be an important factor in the development of the cave system (Glover 1974) and the band may exert a similar control in the West Kingsdale area (Glover 1976).

5. The pre-Carboniferous basement

The most striking difference between Chapel-le-Dale and Kingsdale is the number of risings in each valley. Chapel-le-Dale has five major risings and over 30 minor ones, whereas Kingsdale has one major rising and about five minor risings, depending on water conditions. This contrast stems solely from the disposition of the basement rocks which form the floor of Chapel-le-Dale for 3½ km, whereas no basement rocks are exposed in the floor of Kingsdale.

The presence of the basement not only ensures that several risings are thrown out along the unconformity (Fig 4.2A) but also its relative altitudes affect resurgence location. All early workers emphasized "... the highly irregular nature of the platforms of the pre-Carboniferous surface towards its southern edge..." (Hemingway 1967), and it is now generally accepted that "... at the beginning of the Carboniferous, Yorkshire was an upland area, of fairly substantial relief"(Ramsbottom 1974). The influence of this buried topography on the location of resurgences was first recognised with reference to the Austwick-Horton area (Wilcockson 1927), where drainage from sinks to the east of Gaping Gill re-emerged. Also, in Chapel-le-Dale and Clapham areas "... the position of the majority of resurgences in this area is determined by the topography of the pre-Carboniferous basement rocks and classic examples of this controlling factor are seen at Moses Well, Cat Holes, Skirwith, White Scar Cave, Granite Quarry risings and God's Bridge ..." (Brook 1974). Clearly, the implications of the early suggestion that certain sub-surface watersheds might result from basement influences (Carter & Dwerryhouse 1905), that cave passages might tend to develop along such topographic lows, deserves consideration. It is already established that, from their upper ends, many of the caves around Ingleborough consist of vertical shafts which lead large volumes of water almost through the full depth of the limestone, to a zone close to the basement. Seven caves, in their lowest reaches, have been examined to test the hypothesis of basement control.

Skirwith cave runs in an approximately straight line for almost 1 km, along a line more or less parallel with the geological boundary between the limestone and the basement. On average, the cave is about 100 m from this geological boundary, but since it is as much as 30 m lower in altitude, it must be flowing in a trough in the basement. Like Skirwith, Dale Barn Cave also runs for nearly 1 km in a near-straight line and the original entrance is also situated in a trough in the basement. However, the entrance is 20 m above the lowest part of the cave and only acts as a rising in times of high flow. Thus, this is a case in which the flood overflow is positioned by the basement topography. The present-day normal flow appears to overflow into the adjacent, parallel trough behind Dry Gill. In Dry Gill Cave itself, granular sand contains a high percentage of green rock particles (Long 1970) and thin section analysis of green pebbles taken from the cave proved these to be Ingletonian. These semi-angular pebbles provide evidence both of the mechanical erosion of the basement and of the control of the cave stream by the presence of the basement. Meregill Skit (Fig 4.2A) is approximately 50 m north-east of God's Bridge and is at approximately the same altitude. Whilst God's Bridge is very close to the base of the limestone, at Meregill Skit a 20 m deep flooded shaft was dived in limestone and led to a flooded passage running southeastwards. This passage is kept flooded by the water ponded behind the basement ridge which is cut by the present-day topography at God's Bridge. In the case of Quaking Pot, which appears to approach within 400 m of White Scar Cave, having dropped rapidly to a depth of 143 m, it does not join the White Scar streamway. This non-connection is taken to imply that a continuation of the basement ridge, seen in Chapel-le-Dale between the White Scar resurgence and the resurgence for the Quaking Pot water, acts as an underground watershed (Carter & Dwerryhouse 1905; Ramsden 1974). In the case of White Scar itself, much of the main streamway is within the 'dirty limestones' of the basal beds of the Great Scar and an exposure of Ingletonian rocks in the bed of the streamway has recently been confirmed. In Gavel Pot, both the upstream and downstream sumps are at an altitude of 214.5 m, 1 m above the resurgence level at Leck Beck Head (Bowser 1973). However, divers in the upstream sump found that the water rises up a flooded pot more than 50 m deep (Statham 1975). A possible explanation of this upwelling, and that seen in Deep Rising in Kingsdale, is that the water is being ponded behind a basement ridge.

THE INFLUENCE OF GEOLOGICAL FACTORS ON PROCESSES OF LIMESTONE SOLUTION

1. The role of joints and faults

The solutional enlargement along the joints of bare limestone pavements is one of the most striking features in the landscape of north-west Yorkshire. However, such pavements are nearly always close to a vertical scar and the role of pressure-release of the joints may be significant. Below ground it is difficult to identify points where joints are favouring accelerated solution. The high concentrations of calcium carbonate in the inlets to Swinsto Crawl at shallow depth reinforces the suggestion that, where precipitation falls on to limestone strata, much of the solutional activity takes place at, or close to, the soil bedrock boundary. In this instance the inlets average 138 ppm $CaCO_3$ compared with the main stream sink of 65 ppm.

In the case of the Craven Fault zone, there can be little doubt that the presence of faults is a major control on the chemistry of waters rising within this zone. Compared with the risings in Chapel-le-Dale, those along the fault line demonstrate considerably higher water hardness. The water contains more calcium, magnesium, bicarbonate, sodium and potassium and the variations in hardness are smaller. The mean magnesium hardness of the sites in the group was 32 ppm, compared to an average for all other sampling sites of 15 ppm. This indicates that there is an unusual amount of magnesium bicarbonate being taken into solution, which would seem to indicate that dolomitization has occurred along the fault lines. As dolomite is less soluble than calcite, this feature would, in turn, account for the notably fewer caves developed in this zone.

2. Insolubility of impermeable headstream areas

The geology of the headstream catchment area can directly influence the chemical characteristics of the water flowing off on to the limestone. Values of calcium hardness for water drainage from the base of the Millstone Grit and from peat bogs, at 6.5 ppm (Richardson *personal communication*) and 5-6 ppm, respectively, are close to that in rainwater where the trace is attributed to dust in the atmosphere. However, where water has drained through Yoredale strata, it may already contain significant amounts of calcium in solution before it sinks into the main limestone mass. Examples include Fell Beck (Pitty 1974), Winterscales Beck, and Long Gill in Kingsdale, where the calcium hardnesses are 48, 39 and 43 ppm, respectively.

In addition to direct effects on the character of the water flowing on to the main limestone mass, the influence of the impermeable caprock is usually distinctly discernible in the water when it rises at the base of the limestone. Risings, which are resurgences of water from sink holes, have also been described as conduit-flow risings because the water flowing to them through the limestone tends to follow well-defined conduits or channels resulting from localization of groundwater flow paths by solutional modification (Shuster & White 1971). These resurgences respond rapidly to precipitation, with an increase in discharge and a decrease in hardness. Thus, these resurgences have a lower mean calcium hardness and higher mean discharges than the generally less variable diffuse-flow risings which have no known feeders. On the eastern side of Chapel-le-Dale, where the risings are mainly conduit-flow type, the mean calcium hardness is 128 ppm and the mean coefficient of variation is 17.5 per cent. If only known sink-fed risings are included, the average of means drops to 100 ppm and the coefficient of variation increases to 23.6 per cent. In contrast, on the western side of Chapel-le-Dale, where the risings are almost entirely diffuse-flow fed, the mean calcium hardness is 154 ppm and the mean coefficient of variation 11.3 per cent.

It is important that both discharge and hardness react to precipitation in identifying resurgences. For instance, the Chapel-le-Dale sampling revealed that Dry Gill Cave responded rapidly to precipitation with an appreciable rise in discharge but with no corresponding decrease in hardness. This is believed to result from rapid drainage of the bare limestone area of the catchment, expelling water already held within the limestone, although not freely draining under the influence of gravity alone. Thus the rising, in spite of its rapid discharge responses to precipitation, is really a diffuse-flow rising, but one with a large underground reservoir

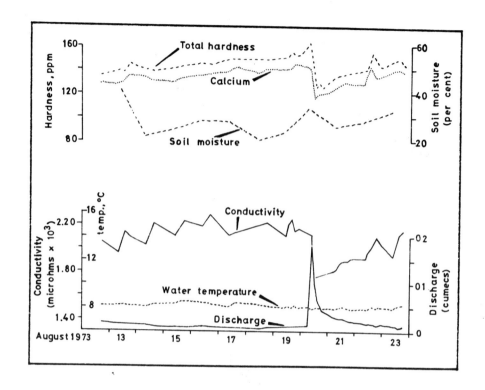

Figure 4.4 Effect of heavy rainfall, after prolonged dry spell, on some chemical and physical characteristics of flow from the Granite Quarry risings, Ingleton. ("Granite" is a local trade-name for the Ingletonian quarried).

behind point of emergence. This reservoir has, in fact, been proved by cave divers who have entered the flooded passages (Statham 1976).

Resurgences are not only typified by greater variability in solute concentration, on average, compared with diffuse-flow risings. During single storm events, water from surface sinks can dilute the concentration in the base-flow within a few hours. The rate of such responses to precipitation in a cave system tends to be more rapid than with a surface river because its conduit feeders are unable to 'overflow their banks' during high water. For example, in a flood observed (Fig 4.4) at the Engine Shed risings in Chapel-le-Dale, the solute concentration of calcium dropped by 23 per cent within six hours, with both calcium concentration and discharge then taking 75 hours to return to their pre-flood base levels (Cavanagh 1974).

3. Impermeability and topography of the pre-Carboniferous basement

Whilst it is obvious that the basement rocks exert little direct influence on solute concentrations or loads, it has been found that the basement exerts a strong indirect

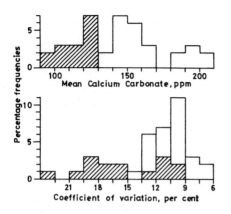

Figure 4.5
Influence of the 'fossil' relief of the pre-Carboniferous landsurface on the solute characteristics of contemporary risings. Risings located on 'highs' of the fossil relief (blanks) are compared with the lower means and greater variability of those risings at 'lows' (shaded)

control. The results of water sampling in Chapel-le-Dale suggested that the risings could be split into two groups. The first group has a high calcium solute concentration (mean 166 ppm), low hardness variability (mean coefficient of variability 9.4 per cent), low discharge (usually less than 1 litre/sec) and low discharge variation. The second group is characterised by a lower calcium hardness (mean 114 ppm), high hardness variability (mean coefficient of variability of 17.7 per cent), high discharge (up to 400 litres/sec) and a high variation in discharge. These groupings have characteristics similar to those suggested by Shuster & White (1971) and applied by Ternan (1972) to the Fountains Fell and the Malham Moor area. When this grouping had been made in the Ingleborough area, it became obvious that the two groups were also differentiated by their position in relation to the topography of the impermeable basement. When the risings were split into those in basement troughs and those on ridges in the basement, and each of these groups split into those with above- or below-average calcium hardnesses (Fig 4.5), noteworthy differences were revealed between the groups. The high hardness group risings, the diffuse-flow risings of Shuster & White, were nearly all situated on high points on the basement ridges. In contrast, the majority of low hardness group risings, the resurgences, were located in depressions in the basement. It has already been demonstrated that the basement lows locate the position of resurgences. Water chemistry shows that not only are these the lines of concentrated flow of high volume, but that the solute concentration is also low. The development of caves along these lows, by both mechanical and chemical erosion, is likely to be self-accentuating.

GROUPING OF THE RISINGS ACCORDING TO SOLUTE CHARACTERISTICS

It is possible to identify most of the influences examined above in the results of water sampling from the Ingleborough area (Fig 4.5). Using the variability of calcium, alkaline and magnesium content of the waters, together with their temperatures and geographical location, it is possible to

Solutional processes in N.W. Yorkshire

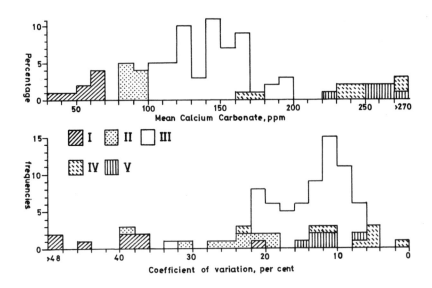

Figure 4.6   Calcium carbonate means and variability, according to the five provisionally identified groups described in the text

identify five groups of sites. Each of these groups represents a different expression of the interplay of these factors.

The first group of sites is obviously that of points where the water can be sampled before it sinks into the main limestone body, such as Kingsdale Beck or Winterscales Beck. This group of sites, which in general have a lower mean hardness (mean $CaCO_3$ = 56.4 ppm) and greater variability of hardness (mean coefficient of variation = 40.2 per cent), is clearly distinct from the large surface outflows leaving the limestone, such as Clapham or Austwick Beck. The calcium and alkaline hardnesses of these 'pre-limestone' waters varies more independently than at many of the other possible sampling sites in the study area, perhaps because of the more marked influence of non-carbon dioxide controlled reactions causing calcium to go into solution, such as the sulphuric acids from peatland drainage. Being surface streams, this group of sites inevitably shows the largest range (mean range = 13.0^0C) and variability of water temperatures in the study area.

The second group is equally readily identified as that of the large risings, such as God's Bridge, White Scar Cave, and other sites where the location is strongly influenced by the basement rocks. This group has mean hardnesses (mean $CaCO_3$ = 89.4 ppm) which do not greatly exceed that of the surface sites but the variations in hardness are less extreme (mean coefficient of variation = 25.0 per cent). The variations which do occur result from the rapid flowthrough of flood pulses which lower rapidly the solute concentrations from their seasonal norm. The calcium and alkaline hardness variations at these sites tended to be more variable than at the smaller risings due to the shorter residence times of the water underground. The variations in water temperature (mean annual range = 6.4^0C) are inevitably significantly less than at the surface stream sites.

The third group which can be identified are the permanent but small-flow risings which form the majority of the risings in the study area. Like the major risings, they occur close to the base of the limestone and often at low points on the unconformity. In addition to showing the most consistent discharges, these sites also reveal the most consistent water hardnesses and temperatures, both of which were distinct from that for the area as a whole. The mean calcium hardness at these sites varies between 100 ppm and 200 ppm, and the group may be subdivided according to the coefficient of variation of calcium hardness. The 15 risings around the edge of Scales Moor which have coefficients of variation for calcium hardness of less than 10 per cent are presumed to be diffuse-flow only risings, with the low mean hardness perhaps reflecting the lack of soil cover over much of Scales Moor. At most of the sites in this group as a whole, the calcium hardness is always in excess of the alkaline hardness. The narrow temperature range results from a lack of response to heavy rainfall (or snowmelt) which produces dilution pulses of cold water in winter and relatively warm water in summer.

The fourth group which can be identified consists of many of the risings in the Craven Fault zone between Ingleton and Austwick. These sites demonstrate high calcium (mean $CaCO_3$ for all sites exceeds 200 ppm), alkaline and magnesium hardness, together with high sodium and potassium contents at some sites. They may be distinguished from the third group of sites because they do not occur along an exposure of the unconformity, they have much higher magnesium contents (mean magnesium hardness = 32 ppm) and, on many sampling occasions, the alkaline hardness was found to exceed the calcium hardness. These sites are thought to be predominantly an expression of the geohydrological influence of the Craven Faults.

The fifth group of sites are similar in many ways to the fourth group, including sites along the Old Road from Clapham to Ingleton and the two sites near Twistleton Hall. Like the fourth group of sites, all reveal high mean calcium, alkaline and magnesium hardnesses together with high mean sodium and potassium contents. However, both discharge and hardness fluctuate considerably at these sites and this separates them from the fourth group of sites. These sites are probably fed by local soil drainage with the high hardnesses resulting from solution at the soil-bedrock interface and the high sodium and potassium contents resulting from salt licks and animal manure.

In addition to the various groups suggested above, the widespread water sampling across the area also spotlighted a number of unusual sites which invite further investigation. Ellerkeld is one obvious example, being the only site not in the Craven Fault zone where the annual mean alkaline

Solutional processes in N.W. Yorkshire

hardness exceeded the annual mean calcium hardness. It is noteworthy that the only site where water was sampled from an artificial conduit, the Yarlsber Farm borehole, has a very high magnesium content. Dry Gill was found to be unusual because solute concentrations hardly varied whilst discharge responded dramatically to heavy rainfall. Skirwith Cave has all the characteristics of a diffuse-flow fed site but is subject to large-scale short-term variations in discharge and solute concentration. This was interpreted as indicating that the cave was being used as a flood over-flow by the stream in Crina Bottom, as subsequently confirmed by dye tests.

CONCLUSION

The influence of physical controls, both within and surrounding the limestone mass to the west and south of Ingleborough, has been shown to affect directly cave development and thus water flow, speed, and direction. They also influence the solute types and concentrations at the risings draining the limestone. Their influence has also been shown to vary with both geological, and to a lesser degree, shorter time spans. It has been shown that by a geographically comprehensive study of an area of marked but consistent geological contrasts, the factors accounting for differences in solutional forms and processes can be identified.

APPENDIX

1. <u>Localities</u>

A list of localities mentioned in the above account (Table 4.2) can be used in conjunction with the Ordnance Survey maps, either the 1:50 000 *Wensleydale and Wharfedale* Sheet 98, or the 1:25 000 *The Three Peaks*. However, it must be stressed that many of these localities are on private land to which permission for access must be sought. In addition, it is essential that no underground exploration be attempted without experienced cavers being in the party. Addresses of caving clubs are obtainable from the National Caving Association, c/o Department of Geography, The University, Birmingham B15 2TT

2. <u>Cave exploration and survey literature</u>

Despite being prolific and apparently widely scattered, there is a comprehensive collection of 'underground literature', in the literal sense, in the Reference Section of the Public Library in Matlock, Derbyshire.

Table 4.2

Alum Pot	SD 775756	Juniper Gulf	SD 765733
Austwick	SD 7668	Kingsdale	SD 6976
Birks Fell Cave	SD 931770	Leck Beck Head	SD 660800
Borrins Moor Cave	SD 770752	Long Gill (Kingsdale)	SD 7180
Brow Gill-Calf Holes	SD 800777	Manchester Hole	SE 100764
Cat Hole	SD 748700	Meregill cave system	SD 740757
Chapel-le-Dale	SD 7377	Moses Well	SD 751702
Clapham	SD 7469	Nick Pot	SD 770737
Dale Barn Cave	SD 711756	Nidd Heads	SE 105732
Deep Rising (within		Quaking Pot	SD 731740
West Kingsdale System)	SD 6977	Rat Hole Inlet	SD 751727
Dent	SD 7087	Rift Pot	SD 761729
Devis Hole Mine Caves	SE 052961	Simpson's Caves	SD 695777
Dowber Gill Cave		Simpson's Pot	SD 696779
(within Dow Cave)	SD 985743	Skirwith Cave	SD 709738
Dry Gill Cave	SD 719759	Smeltmill Beck Cave	NY 849146
Ellerkeld	SD 747743	Spectacle Pot	SD 711777
Engine Shed rising	SD 720752	Storrs Cave	SD 703732
Fell Beck	SD 7573	Swinsto Hole	SD 694775
Fountains Fell	SD 8671	Tatham Wife Hole	SD 732747
Gaping Gill	SD 751727	Turn Dub	SD 797749
Gavel Pot	SD 666791	Twistleton Hall	SD 701751
God's Bridge	SD 732764	Ullet Gill	SD 723763
Goyden Pot	SE 100762	Washfold Pot	SD 774765
Granite Quarry risings	SD 720751	Weathercote	SD 739775
Grid Pot	SD 669847	White Scar Cave	SD 712745
Growling Hole	SD 712774	Windegg Mine Caverns	NZ 012052
Horton	SD 8172	Winterscales Beck	SD 7479
Ingleborough	SD 7474	Yarlsberg Farm	SD 706722
Ingleton	SD 6973		

REFERENCES

Atkinson, F., 1949. The Cavern, Ireby Fell, Lancashire. *Cave Science*, 9, 21-27

Bott, M.H.P., 1967. Geophysical investigations of the Northern Pennine basement rocks. *Proceedings of the Yorkshire Geological Society*, 36, 139-168

Bowser, R., 1973. The Three Counties System. *London University Caving Clubs Journal*, 14, 8-13

Brook, D., 1974. The caves of Chapel-le-Dale and Newby Moss. In: *Limestones and caves of north-west England*, ed. A. C. Waltham (David & Charles, Newton Abbott), 335-342

Carter, W.L. & Dwerryhouse, A.R., 1905. The underground waters of N.W. Yorkshire. Part 2: The underground waters of Ingleborough. *Proceedings of the Yorkshire Geological Society*, 15, 248-304

Cavanagh, A.H., 1974. *Short-term variations in limestone hydrology, Chapel-le-Dale, north-west Yorkshire.* Unpublished BSc dissertation, University of Hull

Coase, A., 1977. The Cave System (Dan-Yr-Ogof). *Transactions of the British Cave Research Association*, 4, 295-324

Dakyns, J.R., Tiddenham, R.H., Gunn, W. & Strahan, A., 1890. The geology of the country around Ingleborough. *Memoir of The Geology Survey of the United Kingdom* (Her Majesty's Stationery Office, London)

Doughty, P.S., 1968. Joint densities and their relation to lithology in the Great Scar Limestone. *Proceedings of the Yorkshire Geological Society*, 36, 479-512

Dunham, K.C., Hemingway, J.E., Versey, H.C. & Wilcockson, W.H., 1953. A guide to the geology of the district round Ingleborough. *Proceedings of the Yorkshire Geological Society*, 29, 77-115

Dunham, K.C., 1974. Granite beneath the Pennines in North Yorkshire. *Proceedings of the Yorkshire Geological Society*, 40, 191-194

Garwood, E.J. & Goodyear, E., 1924. The Lower Carboniferous succession in the Settle district and along the line of the Craven Faults. *Quarterly Journal of the Geological Society*, 80, 184-273

Gascoyne, M., 1973. The caves and potholes of East Kingsdale and Scales Moor. *Lancaster University Speleological Society Journal*, 3, 11-58

Glover, R.R., 1974. Gaping Gill - some underground controls of development. *Craven Pothole Club Journal*, 5, 58-65

Glover, R.R., 1976. The *Porcellaneous* band in Yorkshire caves - Part I. *Bulletin of the British Cave Research Association*, 12, 11-12

Grainger, B.M., 1938. Survey and report on geological structure of "Hensler's Passage Gaping Ghyll, Yorkshire". *Caves and Caving*, 1, 110-112

Halliwell, R.A., 1974. A history of karst studies in Yorkshire. *Transactions of the British Cave Research Association*, 1, 223-230

Halliwell, R.A., 1977. *Aspects of limestone waters near Ingleton, North Yorkshire.* Unpublished PhD thesis, University of Hull

Hemingway, J.E., 1967. The sub-Carboniferous basement in northern England - discussion. *Proceedings of the Yorkshire Geological Society*, 36, 241-244

King, C.A.M., 1960. *The Yorkshire Dales.* Geographical Association

Long, M.H., 1970. Dry Gill Cave. *Bulletin of the British Speleological Association (New Series)*, 1, 1-6

Myers, J.O., 1948. The formation of Yorkshire caves and potholes. *Transactions of the Cave Research Group of Great Britain*, 1, 26-29

O'Nions, R.K. et al, 1973. New isotopic and stratigraphical evidence on the age of the Ingletonian: probable Cambrian of Northern England. *Quarterly Journal of the Geological Society of London*, 129, 445-452

Palmer, A., 1975. The origin of maze caves. *Bulletin of the National Speleological Society of America*, 37, 56-57

Pitty, A.F., 1974. Karst water studies in and around Ingleborough Cavern. In: *Limestone and caves of north-west England*, ed. A. C. Waltham (David & Charles, Newton Abbot), 127-139

Ramsbottom, W.H.C., 1974. Carboniferous - Dinantian. In: *The geology and mineral resources of Yorkshire*. ed. D. H. Rayner & J. E. Hemingway (Yorkshire Geological Society), 47-73

Ramsden, R.W., 1974. The Quaking Pot extension. *Bulletin of the Cave Research Association*, 6, 23-26

Rule, A., 1910. Gaping Ghyll: exploration and survey: Spout and Rat Hole. *Yorkshire Ramblers Club Journal*, 3, 186-192

Ryder, P.F., 1975. Phreatic network caves in Swaledale, Yorkshire. *Transactions of the British Cave Research Association*, 2, 177-192

Shuster, E.T. & White, W.B., 1971. Seasonal fluctuations in the chemistry of limestone springs: a possible means for characterising carbonate aquifers. *Journal of Hydrology*, 14, 93-128

Simpson, E., 1935. Notes on the formation of Yorkshire caves and potholes. *Proceedings of the University of Bristol Spelaeological Society*, 4, 224-232

Statham, O.W., 1975. Gavel Pot. *Cave Diving Group Newsletter*, 34, 13

Statham, O.W., 1976. Dry Gill. *Cave Diving Group Newsletter*, 41, 11-12

Sweeting, M.M. & Sweeting, G.S., 1969. Some aspects of the Carboniferous limestone in relation to its landforms. *Revue de Géographie des pays Méditerranéens*, 7, 201-209

Ternan, J.L., 1972. Comments on the use of a calcium hardness variability index in the study of carbonate aquifers: with reference to the Central Pennines, England. *Journal of Hydrology*, 16, 317-321

Wager, L.R., 1931. Jointing in the Great Scar limestone of Craven and its relation to the tectonics of the area. *Quarterly Journal of the Geological Society of London*, 87, 392-424

Waltham, A.C., 1970. Cave development in the limestone of the Ingleborough district. *Geographical Journal*, 136, 574-585

Waltham, A.C., 1971. Shale units in the Great Scar limestone of the southern Askrigg block. *Proceedings of the Yorkshire Geological Society*, 38, 285-292

Wilcockson, W.H., 1927. The pre-Carboniferous topography near Austwick and Horton-in-Ribblesdale. *Proceedings of the Yorkshire Geological Society*, 21, 17-23

CHAPTER 5   RUNOFF PROCESSES IN TROPICAL RAINFORESTS WITH SPECIAL REFERENCE TO A STUDY IN NORTH-EAST AUSTRALIA

D. A. Gilmour[*] &   M. Bonell[**]

INTRODUCTION

The spatial distribution of overland flow (Horton 1933, 1945) may be more restricted than originally envisaged, and the 'variable source area' concept (Hewlett & Hibbert 1967; Hewlett & Nutter 1970; Hewlett 1974) has become increasingly accepted in humid temperate forest areas to explain the source of storm runoff or 'quickflow' (Dunne et al 1975). However, it is uncertain whether this concept applies similarly in tropical rainforest. At present there is little quantitative information on natural runoff processes under mature, tropical rainforest (Thomas 1974) with most work confined to broad-scale water balance studies, as in Malaya (Kenworthy 1969; Low & Goh 1972) and West Africa (Ledger 1975). However, the hydrological effects of a change in land-use from rainforest to tea and pine plantations have been measured in detail in selected catchments in East Africa (Blackie 1964; Dagg & Blackie 1965; Blackie 1972), and the results of this work summarized (Pereira 1973). Geomorphologists frequently comment on surface runoff in rainforest areas in relation to landscape denudation, but their speculations are usually based on qualitative observations (Douglas 1969, 1973; Ruxton 1967; Thomas 1974; Tricart 1972).

The purpose of the present contribution is to present details of measurements made near Babinda, in north-east Queensland (Fig 5.1) in order to augment the sparse data on actual hydrological processes which operate on the rainforest floors in the tropics.

THE RAINFOREST ENVIRONMENT

1. Climate

A distinctive feature of temperatures in tropical climates, apart from the continuous warmth, is that diurnal usually exceed mean monthly ranges (Whitmore 1975). In forests close to the equator, the heavy monthly rainfall totals also exhibit little variation. Away from the equator seasonally dry climates are recognisable, as in much of north-east Australia, with droughts in most years. Rainfall is much more intense than in temperate regions, with about 40 per cent falling at rates of 25 mm/hr or more. Only 5 per cent of the precipitation in mid-latitudes falls at such

[*] Forestry Training & Conference Centre, Gympie, Queensland
[**] Dept of Geography, James Cook University of North Queesland

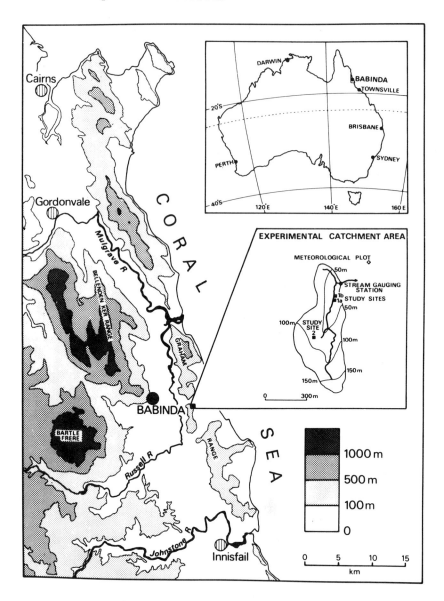

Figure 5.1 The location of study area and experimental plots in South Creek

intensities (Hudson 1971). Thus the effect of more frequent, heavier, downpours is that the kinetic energy of rain falling at intensities greater than 25 mm/hr in the tropics is about 16 times greater than that of temperate rainfall. In addition, tropical rainfall also owes its higher kinetic energy to the increase in drop size with increasing intensity, although this trend is reversed at very high intensities (Hudson 1971). Intensity is associated particularly with tropical cyclones or an active inter-tropical convergence zone (ITCZ). Maximum short-term intensities of an hour or less are generally associated with thunderstorms in the moist air of the equatorial trough itself. However, in areas $5°$ or more from the equator, highly organised disturbances develop due to the increase in the Coriolis force. Such tropical cyclones produce the highest absolute totals for storms in the tropics, particularly for falls lasting more than 6 hours (Jackson 1977). In consequence, localities within $5°$ of the equator have less concentrated rainfalls than those in the cyclone-prone belt. For instance, the daily rainfall likely to be equalled or exceeded once every 100 years is less than 250 mm in the vicinity of Kuala Lumpur. In contrast the Babinda area, together with much of north-east Queensland, regularly experiences daily falls of over 250 mm and the 100 year recurrence storm is likely to equal or to exceed 700 mm (Lockwood 1974).

2. Vegetation

 a) *Plant communities*. The tropical rainforest is the most complex of all plant communities (Webb & Tracey 1973) since it represents vegetation favoured by radiation, moisture, nutrients, shelter, and soil aeration at near-optimum values. However, since the rainforest encompasses a wide variety of vegetation complexes and a wide range of environmental conditions, it is a difficult plant community to define comprehensively. In general, it is a closed community of broad-leaved, evergreen moisture-loving trees, usually with two or more layers of trees and shrubs present. Stem-forms are usually distinctive and often curious in shape, and vines and epiphytes are frequently present (Schimper 1903).

 b) *Productivity and litterfall*. Compared with other forested regions of the world, the plentiful moisture supply together with high temperatures and a long growing season, including high insolation rates during photosynthesis, account for the high productivity of humid tropical rainforests. Productivity is ten times that of arctic-alpine environments, more than three times that of cool temperate forests and twice that of warm temperate forests (Bray & Gorham 1964). The effect of high productivity on tropical forest floor hydrology relates to characteristics of litterfall and decomposition. An average figure for litter production in equatorial forests is 10.9 tonnes/ha/yr (Bray & Gorham 1964). Specific examples include that of 7.2 - 13.4 tonnes/ha/yr for the Ivory Coast (Bernhard 1970) and 7.3 tonnes/ha/yr in the Amazon forest (Klinge & Rodrigues 1968). Bailey (1976) quotes figures of 7.2 - 9.3 tonnes/ha/yr for north-east Queensland rainforest and Brasell *(personal communication)* measured amounts of 8.7 - 9.7 tonnes/ha/yr for the same region. Decomposition rates, however, match those of productivity, with organic matter on rainforest floors in Java, for instance, decomposing at a rate of 8.1 tonnes/ha/yr and that in lowland forests in Borneo at 10.7 tonnes/ha/yr (Wanner 1970).

c) *Influence of forest cover on kinetic energy of rainfall*. Research in temperate pine forest suggests that a canopy re-distributes a volume of rain among drops of all sizes, irrespective of rainfall intensity (Chapman 1948), and that droplets are sufficiently enlarged during interception in the canopy to attain terminal velocity within a 9 m fall after release. In tropical forests, the structure of the understorey is therefore the critical control rather than canopy height.

The consequence of high intensity rainfall or throughfall is the possibility of the infiltration capacity of the forest floor being exceeded (Hillel 1971). For instance, an increase in overland flow from the savanna to the humid tropical forests of the Kanuku Mountains of Guyana has been attributed to the greater amount of precipitation, exceeding 3600 mm annually, in the mountains and to its greater intensity (Kesel 1977). In New Guinea, overland flow is favoured where the canopy is incomplete, due to treefall resulting in greater exposure of the surface soil to raindrop compaction (Ruxton 1967). Part of the rainfall intercepted in the forest canopy may reach the ground as stemflow, and such flows representing as much as 15 per cent of the total precipitation have been estimated (Rutter 1963). In contrast, storms of 40 - 65 mm rainfall in a rainforest stand in Tanzania yielded only 0.9 mm of stemflow (Jackson 1971). However, even small yields of stemflow could be quite important in generating runoff since it is concentrated on to a small zone round the base of tree trunks with the possibility of localised saturation (Ruxton 1967; Douglas 1973).

3. Soils

Tropical rainforest soils are usually the Oxisols of the humid tropical climates with base and general nutrient deficiencies resulting from vigorous leaching under the high temperatures and rainfall which prevail. Although organic content is usually low, comparatively large amounts of nutrients are tied up mainly in the organic matter. In New Guinea, under closed canopy forest, leaf litter may average less than 2 cm, even on gentle slopes, with thicknesses of less than 0.3 cm on slopes steeper than 35 degrees (Ruxton 1967). In the mineral fraction of rainforest soils, the widely recognised 'lateritic' characteristics include a red or yellow clay fraction enriched in alumina and iron oxides but depleted in silica. Although generally loamy or clayey in texture, upper horizons are often sandy (Haig *et al* 1958). In fact, the essential edaphic factor favouring rainforest growth is that soils should be relatively permeable and well-drained (Isbell 1973). For example, much of the Amazon forest stands on inherently nutrient-deficient alluvium (Stark 1971) and in deeply weathered gneiss profiles near Ibadan, infiltration rates are exceedingly high (Nye 1955). However, it is easily appreciated that, on sloping ground such sandy soils may be particularly susceptible to splash erosion. Further if particles detached by splash block soil pores, overland flow might be anticipated as infiltration rates decline.

The rapid rate of organic matter decomposition is also significant, since it leaves a sparse litter cover on the forest floor. The activity of decomposer organism is favoured by the high temperatures in conjunction with sustained high moisture conditions to a greater degree than in the equivalent cleared rainforest or savanna areas (Madge 1965, 1969). For example, 40 - 50 per cent of the litter is decomposed within 5 weeks on the floor of mature forests in Guatemala (Ewel 1976), and the elements released may be rapidly lost in drainage water. In the Amazon forests, the first flush of runoff appears to leach away the bulk of available nutrients (Jordan & Kline 1972).

In the Babinda area, the rainforest floor includes much bare soil and tree roots of all sizes commonly occurring on or above the soil surface. Where roots grow across the slopes, the upslope side accumulates downwash of litter and soil whereas the lower side is often eroded away into a small step. The main characteristics of the Babinda rainforest are therefore those of rainforests' floors in general, but with the rare scientific advantage of relative ease of access for the instrumentation of the hydropedological processes operating on its floor.

HYDROPEDOLOGICAL INVESTIGATIONS IN THE BABINDA RAINFOREST

1   The physical setting

The study site in the Babinda area is situated on the slopes of a low range which separates the coastal plain from the Coral Sea at latitude $17°20'S$. Immediately west of the narrow coastal plain rises the precipitous massif of the Bellenden Ker Range with Queensland's highest peak, Mt. Bartle Frere (1610 m) at its southern end.

The adjacent high mountain range together with its alignment across the prevailing SE winds ensures a high precipitation on the coast and nearby ranges. The climate of the area is typical of that experienced on the wet tropical coastal belt with high year-round temperatures and humidities and high annual rainfall totals being the outstanding features. The average annual rainfall at the study site is 4194 mm. Cyclones and tropical lows of relatively frequent occurrence are a characteristic and significant feature of the climate and give rise to rainfall intensities which are among the highest recorded in Australia.

The natural vegetation cover of the area is tropical rainforest (Mesophyll Vine Forest according to Webb's (1959) classification). The regular occurrence of cyclones in the region has proved to be a potent force in modifying the composition and structure of the rainforest. The strong winds which accompany many cyclones are responsible for causing most damage which may vary from broken branches to the complete levelling of large areas of rainforest. The forest in the study area strongly reflects the modifying force of cyclones, with a low tree height (25 - 35 m) and a tendency towards small-diameter trees.

Figure 5.2 The forest floor under undisturbed lowland rainforest in the catchment near Babinda, showing the sparsity of leaf litter and the presence of numerous surface roots.

2. Visible indications of modes of flow on the Babinda rainforest floor

There must be a very rapid flow path for precipitation to find its way into the stream channels since the stream rises very quickly during rainfall events. The condition of the forest floor (Fig 5.2), with the sparsity of litter, presence of exposed roots, and root steps on slopes all suggest the importance of direct surface runoff. Equally, rapid subsurface flow would be significant since the high density of tree roots and their rapid decay in the humid tropical environment, together with worm and other biopores maintain high porosities in the surface soils. Also, most first order streams in the catchment gather in clearly defined stream head hollows (Fig 5.3) which probably reflects coalescing lines of subsurface flow. However, as in most visual inspections of hydropedological processes on rainforest floors, the relative importance of the various flowpaths and the mechanisms involved cannot be specified by qualitative observations alone.

In order to record some actual measurements of processes previously described largely in qualitative terms, experiments were set up to monitor hydrological interactions with the soil and their influence on the development of overland flow, slope wash, and streamflow beneath a tropical rainforest. The study area chosen, in the high-rainfall rainforest area of north-east Queensland, is a small gauged rainforest catchment in the foothills of the tropical lowlands near

Figure 5.3 Stream head hollow marking the commencement of a first-order stream in this area of undisturbed rainforest. This example is typical of most first-order streams in the catchment.

Babinda (Figs 5.1 and 5.4). The catchment is 26 ha in area, with steep slopes averaging 19 degrees. There is no saturated floodplain adjacent to the stream in this steeply sloping area. The soils are yellow-brown clays derived from deeply weathered metamorphic rocks.

photo by B. J. Pillans

Figure 5.4   Regional setting of the experimental area. South Creek catchment is shown in the lower left of the photo, with an adjacent research catchment (North Creek) in the lower right. The meteorological station is visible in the centre at the bottom edge. The rainforest-covered granite massif of the Bellenden Ker Range in the background falls steeply to the coastal plain. The small town of Babinda is located in the middle distance, amid sugar cane plantations.

3. Description of meteorological conditions producing rainfall events

Certain typical surface meteorological conditions are established for the summer when heavy rainfall is persistent. Fig 5.5 shows a very active monsoonal trough (ITCZ) which has developed across northern Australia containing tropical lows in the Gulf of Carpentaria and the north-west Coral Sea. This particular Coral Sea low produced a 3-day rainfall total of 461 mm in the Babinda experimental catchment as it tracked south-west, with a daily total of 238 mm on the 6th. The Gulf low later developed into Tropical Cyclone 'Otto', bringing a further 126 mm to the catchment as it degenerated whilst moving south-east. Later in the year, moist on-shore south-easterly winds occur sporadically during the 'dry' season. These rains are of relatively low intensity, although daily totals may still exceed 100 mm. The meteorological system usually responsible on the surface is a firm coastal ridge of high pressure but whether rainfall occurs or not seems to depend on the lower easterlies gaining sufficient depth at the expense of the upper westerlies. The

Runoff in tropical rainforests

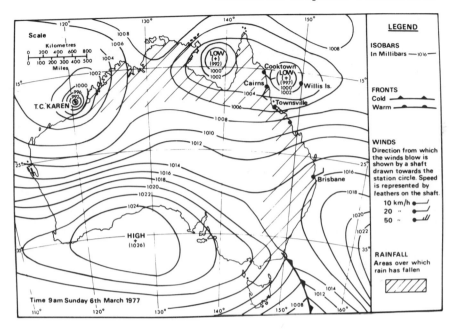

Figure 5.5 Typical example of the surface meteorological chart for the summer when heavy rainfall is persistent.

'wet ridge' situation at surface level (Fig 5.6) produced 32 mm of low intensity rainfall in the Babinda catchment. The difference in intensity between the winter and summer situations can be depicted as a histogram of maximum weekly 6 minute intensities (Table 5.1). The distribution is clearly bimodal with the first two categories representing the 'winter' situation and the remainder the 'summer' situation when 6 - 10 mm 6-minute intensities commonly occur. However, much higher totals have been recorded such as in the exceptionally heavy storm of December 20, 1976, described elsewhere (Gilmour & Bonell 1979).

4. Instrumentation

The experimental work was designed to trace the movement of water into, through, and out of the soil profiles. Three sites were selected for study, two on the lower slopes adjacent to the main stream and one on the upper slopes in the area where the first order streams are initiated (Fig 5.1). At each site a bank of four metal troughs was installed into the face of a trench so that lateral flow was intercepted at levels ranging from the surface to 1 m, with intermediary levels at 0.25 m and 0.50 m. The troughs were two metres in length and the collected runoff water from each trough was led into tipping bucket devices and the number of tips recorded on paper tape in a digital event recorder by using mercoid switches attached to the tipping buckets (Fig 5.7). This instrument records on a time basis of six minutes, thus allowing an accurate study to be carried out relating the

Maximum intensity class boundaries, mm	1.6-3.6	3.6-5.6	5.6-7.6	7.6-9.6	9.6-11.6
Frequency	7	2	8	3	1

Table 5.1 Number of maximum 6-minute rainfall intensities per week for a given intensity class (from 14th January to 16th June 1976).

onset of rainfall to the onset of lateral flow from the various soil horizons. Thus the nature of the recording system means that a large number of observations for individual storms can be collected which offsets the small statistical sample of plot studies in the catchment. The major problem with using troughs is that they are capable of distorting subsurface flow patterns. It is known that the soil-water-air interface at the trench face produces an impediment to flow. This acts in such a way as to prevent the emergence of water until a substantial positive head has built up at the back of the trench. The end result is that soil water by-passes the trench face and therefore distorts the results by giving the misleading impression that no water, or only small amounts, appear to be moving at that depth (Hewlett *personal communication* 1977). This is the potential situation particularly at the beginning and termination of a storm in the rainforest. On the other hand, when complete saturation occurs the air-water interface is not important and the troughs tend to act as a local sink, rather analogous with wells, which gives rise to a flow pattern towards the pit (Knapp 1973). This was suspected to be the situation certainly at 0.25 m depth in this experiment during the bulk of a storm event. However, two factors counter this objection. A substantial proportion of stormflow was collected by the surface troughs which were obviously not affected by this problem. In addition, the upslope experiment utilised a deep, natural exposure adjoining a first order stream.

In addition to the runoff troughs two of the sites were instrumented with a network of tensiometers and piezometers so that unsaturated and saturated soil conditions could be monitored. Furthermore two deep wells at the upper slope site 2 were driven so that fluctuations of the saturated zone could be monitored.

5. Saturated hydraulic conductivity (k_s)

The rates of water movement at different depths in the soil were established by various methods (Bonell & Gilmour 1978). The mean k_s value of 32.45 m/day for the surface 10 cm is very high and would not impede even the most intense downpour. There is a sharp decline in k_s with depth, although the value for the 11 - 20 cm depth, at a mean of 1.54 m/day or 64 mm/hr, is still high enough to accommodate most rainfall events. The highest rainfall intensities would exceed the percolation rate of this zone. Below 20 cm depth, the mean k_s drops to a value of 0.32 m/day, a mere 13 mm/hr. Rainfall intensity exceeds this value for

Runoff in tropical rainforests

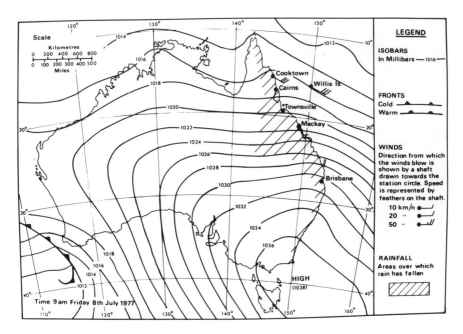

Figure 5.6   An example of a 'wet ridge' situation at the surface during winter

quite long periods on many occasions and water ponds on this impeding layer. Any subsequent rainfall saturates the soil in the top 20 cm so that water leaves the area laterally both through the topmost layers of soil and as surface runoff over most of the catchment. The amount of water flowing laterally through each of the zones depends on a number of factors, the most important of which are the rainfall intensity and duration (Gilmour & Bonell 1977).

6.   The effect of a heavy storm

During the study period, a heavy storm occurred on the 26-27th January 1976. The storm was typical of the intermittent, short-duration falls which are common during the wet season.

   a) The synoptic situation. This rainfall event was attributed to upper air phenomena (Fig 5.8) which were not evident at the surface (Fig 5.9). A surface trough orientated roughly north-west/south-east had moved eastwards in association with the movement of a high pressure ridge for the previous two days. The trough passed through Cairns on the 26th and then stagnated just to the northeast. An upper trough at the 850 - 700 mb. levels moved eastwards in sympathy with the surface trough passing through the Cairns area on the 26th. This trough then retrogressed and by 3 am on the 27th was located in the Babinda area when the storm occurred. Coincident with the retrogression of this trough, an upper level (300 mb.) anticyclone moved eastwards over the area. Thus the strong convective

Figure 5.7 Runoff troughs and tipping bucket gauges used to collect and record the volume of lateral flow from various depths

activity which resulted was due to the low level convergence associated with the trough over the area, re-inforced by upper level divergence corresponding with the passage of a 300 mb. anticyclone.

   b) *Rainfall amount and intensity*. The storm lasted for 5 hours 50 minutes, but the most intense period lasted for only 1 hour. During this hour, 36.72 mm of rain fell, with a maximum intensity of 5.96 mm/6 min. Prior to the storm, the tensiometers showed soil moisture tension values close to zero which is typical of wet season conditions from about January until May or June. At Site 2, overland flow from this storm was very high and greatly exceeded the lateral flow from the 0.25 m horizon. No lateral flow occurred from the deeper zones. There was a close relation between the overland flow graph and the stream hydrograph (Fig 5.10), thus highlighting the importance of overland flow as a major contributor to storm runoff. The short lag response between the peaks of rainfall, overland flow and stream runoff can be statistically verified, using a modified simple lag regression approach which accounts for a high percentage of the total variance (Bonell *et al* 1979). The cross-correlations were demonstrated and it is evident that the peaks between rainfall and overland flow and between the overland flow and stream discharge are both 6 minutes, with a lag of 12 minutes between the overland flow and stream discharge. In the case of rainfall-overland flow, the lag could in fact be much less than 6 minutes. This was suggested by the cross-correlation function where the value at lag zero is almost as high as that for the 6-minute lag.

Runoff in tropical rainforests

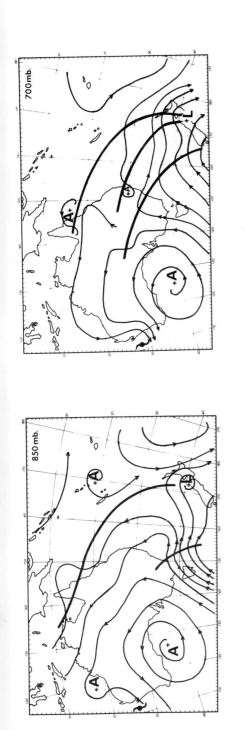
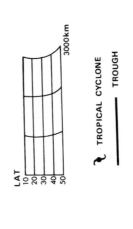
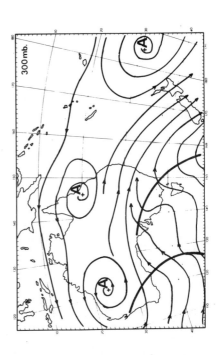

Figure 5.8 Streamline charts for the 850, 700 and 300 mb levels on 27.1.76 (0300 EST)

Rainfall in tropical rainforests

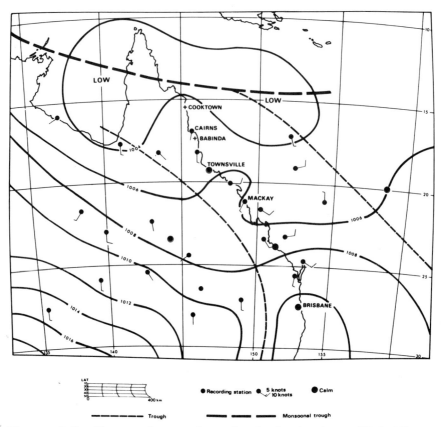

Figure 5.9 The surface meteorological chart on 27.1.76 (0300 EST)

The shortness of these recorded lags is even more extreme since the stream gauging station is located 0.4 km from the experimental site.

The results from the other two sites closely resembled that at Site 2, with the majority of flow occurring as surface runoff. However, there is more flow in the deeper layers, particularly the 0.25 m layer. This is probably due to larger volumes of residual rock throughout the profile creating many macropores. At Site 1a the pattern of storm drainage also remained essentially similar, despite the considerable depth of leaf litter which had covered the site in the wake of tropical cyclone 'Keith' in January 1977. Clearly the present measurements show that leaf litter does not impede overland flow in tropical rainforests, contrary to what has been imagined by many observers.

c) *Subsurface water and flow.* Immediately after rainfall ceases, the widespread overland flow contracts to small rivulets which, in turn, quickly submerge beneath the soil surface. The 0.25 m bucket continues to tip for a short time, reflecting the dissipation of the perched water-table and subsurface flow. It is significant that at the

Runoff in tropical rainforests

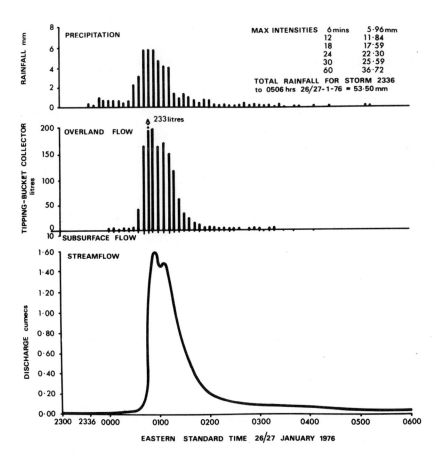

Figure 5.10 Relationships between precipitation, lateral flow from various depths, and stormflow for a sample storm during the 1976 wet season

upslope experimental station, Site 2, the permanent watertable is too deep to initiate saturation overland flow (Bonell & Gilmour 1978). Therefore, although it has been established that a high percentage of the rainfall leaves the catchment during storm events as overland flow, substantial amounts still percolate into deeper water tables, to be subsequently released into the stream as baseflow. Indeed, the large amount of moisture held in the soil profile, amounting to as much as 1.5 m equivalent depth in the top 3 m of soil at the height of the wet season, is probably a significant contributor to the delayed baseflow component of the hydrograph. This is thought to move by the described translatory process (Hewlett & Hibbert 1967) in striking contrast to the abruptness of the hydrograph response to overland flow.

7. Effects of longer-term rainfall totals

Weekly tipping bucket volumes were correlated with fourteen rainfall variables for both summer and winter conditions for the 1976 wet season (Bonell & Gilmour 1978). This seasonal separation was based on the bimodality of the short-term maximum weekly rainfall (Table 5.1). It was found that both short and long term maximum weekly totals are a major control in overland flow generation. A perched watertable is maintained at the soil surface once the critical intensity is attained, which creates a temporary 'impeding' surface during rainfall events. Thus substantial saturation overland flow occurs at this time. This contrasts markedly with the situation in humid temperate areas where, away from areas with a permanent shallow watertable, the longer-term volumes of rainfall are inadequate to maintain temporary saturated conditions and rainfall intensities sufficiently high to exceed prevailing infiltration rates are extremely infrequent. For both these reasons, rainfall intensity is generally not a dominant variable in stormflow generation in humid temperate areas, as observed in forested watersheds in the southern Appalachians (Hewlett et al 1977).

CONCLUSIONS

Many observers have offered different interpretations of the development of overland flow in tropical rainforests. In part, differing opinions reflect the intrinsic variablity of these environments. However, the lack of observational data is equally a source of uncertainty, and the present experiment is one of the first specifying scales and rates of processes which determine the disposition of water after it reaches a rainforest floor. The present observations show that there is a critical decline in permeability of the soil with depth which, in the wet season, has large volumetric moisture contents ranging between 0.4 and 0.5. Evidently, with rainfall intensities frequently higher than subsoil k_s values and in volumes in excess of available pore space, overland flow is a common phenomenon. In addition, it is considered that, during the most intense parts of rainstorms, a substantial proportion of the throughfall fails to enter the soil surface and manages to pass out of the catchment as technically surface storm flow using Hewlett's (1974) terminology. There are some differences in lag response between individual storms which are demonstrated by Bonell et al (1979). This is primarily attributed to the structural differences in the rainfall input which incorporates such aspects as uni- or bi-modality of the rainfall distribution (pulse shape), storm duration and intensity. Antecedent soil moisture is not considered to hold the same degree of significance in the wet season because of the perpetual near saturated conditions.

The overland flow finds its way into the main stream network very quickly, because of the steep catchment slopes and high drainage density, the latter which is, in itself, a response to the regular occurrence of widespread overland flow. This distinctive response accounts for the large contribution or quickflow (or stormflow) to the total volume of streamflow and the associated rapid response of the stream hydrograph. Stormflow in this catchment accounts for 47 per cent of the total annual discharge (Gilmour 1977).

This adjustment between hydrometeorological conditions and channel pattern contrasts with the generally applicable hypotheses for humid temperate, forested areas, where the variable source area concept explains the runoff generating processes. The present investigation suggests, however, that there is a real and fundamental geographical difference between the major world regions although these can be established only by instrumentation and detailed measurement in the field. Even within this particular region it is not claimed that the described processes occur exactly the same way in other lowland rainforest environments. It is intended to extend the current work to check whether extrapolation of the Babinda catchment findings is valid.

ACKNOWLEDGEMENTS

The Queensland Forestry Department and the James Cook University of North Queensland are thanked for their continued support of this work. Additional financial assistance was provided by the Australian Water Resources Council and some of the instruments were made available by the Queensland Irrigation and Water Supply Commission. Field assistance was provided by Mr. H. O. McColl and Mr. M. Devery. Mr. B. Gordon of the Bureau of Meteorology, Brisbane, provided the weather charts which are reproduced with the kind permission of the Regional Director of the Bureau, Mr. W. R. Wilkie.

As much of the writing of the above account was undertaken whilst one of us (MB) was on study leave based at the Department of Geography in the University of Sheffield, we are indebted to Professor R. S. Waters for use of the departmental facilities, to Professor S. Gregory for helpful discussions on tropical climatology, and to Mr. S. Frampton who arranged for the maps to be drawn.

REFERENCES

Bailey, S.W., 1976. Seasonal variation of litterfall in three north Queensland rainforests. *CSIRO Division of Soils Technical Memorandum*, 28/1976

Bernhard, F., 1970. Etude de la litiere et de la contribution au cycle des éléments mineraux en foret ombrophile de Cote-d'Ivoire. *Oecologia*, 5, 247-266

Blackie, J.R., 1964. Hydrology and afforestation in the Aberdares. *East African Geographical Review*, 2, 17-22

Blackie, J.R.,1972. Hydrological effects of a change in land use from rainforest to tea plantation in Kenya. In: *Symposium on the results of research on representative and experimental basins. Publication of the International Association of Scientic Hydrology*, 97, 312-329

Bonell, M. & Gilmour, D.A., 1978. The development of overland flow in a tropical rainforest catchment. *Journal of Hydrology*, 39, 365-382

Bonell, M., Gilmour, D.A. & Sinclair,D.F., 1979. A statistical method for modelling the fate of rainfall in a tropical rainforest catchment. *In preparation*

Bray, J.R. & Gorham, E., 1964. Litter production in forests of the world. *Advances in Ecological Research*, 2, 101-157

Chapman, G., 1948. Size of raindrops and their striking force at the soil surface in a red pine plantation. *Transactions of the American Geophysical Union*, 29, 664-670

Dagg, M. & Blackie, J.R., 1965. Studies of the effects of changes in land use on the hydrological cycle in East Africa by means of experimental catchment areas. *Bulletin of the International Association of Scientific Hydrology*, 4, 63-75

Douglas, I., 1969. The efficiency of humid tropical denudation systems. *Transactions of the Institute of British Geographers*, 46, 1-6

Douglas, I., 1973. Rates of denudation in selected small catchments in Eastern Austrlia. *University of Hull Occasional Papers in Geography*, 21

Dunne, T., Moore, T.R. & Taylor, C.H., 1975. Recognition and prediction of runoff-producing zones in humid regions. *Hydrological Sciences Bulletin*, 20, 305-327

Ewel, J.J., 1976. Litterfall and leaf decomposition in a tropical forest succession in eastern Guatemala. *Journal of Ecology*, 64, 293-308

Gilmour, D.A., 1977. Effect of rainforest logging and clearing on water yield and quality in a high rainfall zone of north-east Queensland. *Institution of Engineers, Australia,* Hydrology Symposium 1977 Brisbane, *National Conference Publication,* 77/5, 156-160

Gilmour, D.A. & Bonell, M, 1977. Streamflow generation processes in a tropical rainforest catchment - a preliminary assessment. *Ibid,* 178-179

Gilmour, D.A. & Bonell, M., 1979. Six-minute rainfall intensity data for an exceptionally heavy tropical rainstorm. *Weather, in press*

Hudson, N., 1971. *Soil conservation* (Batsford, London)

Haig, I.T., Huberman, M.A. & Aung Din, 1958. *Tropical silviculture,* (FAO, Rome), Volume 1

Hewlett, J.D., 1974. Comments on letters relating to "Role of subsurface flow in generating surface runoff; 2. Upstream source areas" by R. Allan Freeze. *Water Resources Research,* 10, 605-607

Hewlett, J.D. & Hibbert, A.R., 1967. Factors affecting the response of small watersheds to precipitation in humid areas. In: *International Symposium on Forest Hydrology,* ed. W. E. Sopper & H. W. Lull, (Pergamon, Oxford), 275-290

Hewlett, J.D. & Nutter, W.L., 1970. The varying source area of streamflow from upland basins. *Proceedings of the Symposium on Interdisciplinary Aspects of Watershed Management,* American Society of Civil Engineers, 65-83

Hillel, D., 1971. *Soil and water* (Academic Press, London and New York)

Horton, R.E., 1933. The role of infiltration in the hydrological cycle. *Transactions of the American Geophysical Union,* 14, 446-460

Horton, R.E., 1945. Erosional development of streams and their drainage basins: hydrophysical approach to quantitative morphology. *Bulletin of the Geological Society of America,* 56, 275-370

Isbell, R.F., 1973. Soils of the tropical and subtropical rainforests of eastern Australia. *Wildlife in Australia,* 10, 70-71

Jackson, I.J., 1971. Problems of throughfall and interception assessment under tropical forest. *Journal of Hydrology,* 12, 234-254

Jackson, I.J., 1977. *Climate, water and agriculture in the tropics.* (Longman, London and New York)

Jordan, C.F. & Kline, J.R., 1972. Mineral cycling: some basic concepts and their application in a tropical rainforest. *Annual Review of Ecology & Systematics,* 3, 33-51

Kenworthy, J.B., 1969. Water balance in the tropical rainforest: a preliminary study in the Ulu Gombak forest reserve. *Malayan Nature Journal,* 22, 129-135

Kesel, R.H., 1977. Slope runoff and denudation in the Rupuruni Savanna, Guyana. *Journal of Tropical Geography,* 44, 33-42

Klinge, H. & Rodrigues, W., 1968. Litter production in an area of Amazonian terra firma forest. *Amazonia,* I, 287-310

Knapp, B.J., 1973. A system for the field measurement of soil water movement. *British Geomorphological Research Group Technical Bulletin,* 9

Ledger, D.L., 1975. The water balance of an exceptionally wet catchment in West Africa. *Journal of Hydrology,* 23, 207-214

Lockwood, J.G., 1974. *World climatology: an environmental approach.* (Arnold, London)

Low, K.S. & Goh, K.C., 1972. The water balance of five catchments in Selangor, West Malaysia. *Journal of Tropical Geography,* 35, 60-66

Madge, D.S., 1965. Leaf fall and litter disappearance in a tropical forest. *Pedobiologia,* 5, 273-288

Madge, D.S., 1969. Litter disappearance in forest and savanna. *Pedobiologia,* 9, 288-299

Nye, P.H., 1955. Some soil forming processes in the humid tropics. Part II: The development of the upper slope member of the catena. *Journal of Soil Science,* 6, 51-62.

Pereira, H.C., 1973. *Land use and water resources in temperate and tropical climates* (Cambridge University Press, Cambridge)

Rutter, A.J., 1963. Studies in the water relations of *Pinus sylvestris* in plantation conditions. I. Measurement of rainfall and interception. *Journal of Ecology,* 51, 191-203.

Ruxton, B.P., 1967. Slopewash under mature rainforest in Northern Papua. In: *Landform studies from Australia and New Zealand,* ed. J. N. Jennings & J. A. Mabbutt. (Australian National University Press, Canberra), 85-94

Schimper, A.F.W., 1903. *Plant geography on a physiological basis,* translated by W.R. Fisher (Clarendon Press, Oxford)

Stark, N., 1971. Nutrient cycling: I. Nutrient distribution in some Amazonian soils. *Tropical Ecology,* 12, 24-50

Thomas, M., 1974. *Tropical geomorphology* (Macmillan, London)

Tricart, J., 1972. *Landforms of the humid tropics, forests and savannas.* (Longman, London)

Wanner, H., 1970. Soil respiration, litter fall and productivity of tropical rainforest. *Journal of Ecology,* 58, 543-547

Webb, L.J., 1959. A physiognomic classification of Australian rainforests. *Journal of Ecology,* 47, 551-570

Webb, J.L. & Tracey, G., 1973. The tropical rainforest in Australia. *Wildlife in Australia,* 10, 66-69

Whitmore, T.C., 1975. *Tropical rainforests in the Far East* (Clarendon Press, Oxford)

CHAPTER 6 EROSION PROCESSES IN SMALL FORESTED CATCHMENTS IN LUXEMBOURG

A. C. Imeson & H. van Zon[*]

INTRODUCTION

Soil erosion has been studied largely by agricultural engineers who are concerned essentially with reducing the erosion risks which may accompany the clearing and cultivation of soils (Hudson 1973; Toy 1977). By comparison, the hazard is much less in forested areas because the tree canopy, litter cover and soil humus protect the soil and because infiltration into forest soils is rapid. Indeed, maximum soil losses from forested catchments are only a few per cent of those recorded for agricultural areas (Smith & Stamey 1965). Nonetheless, forested areas provide critical evidence, even when there is a long history of felling, for establishing erosion rates which could approach the geological norm (Table 6.1). However, the mechanisms which provide a given erosion rate are little studied, despite the unlikelihood of those established as important for agricultural lands being prevalent under forest. The purpose of the present chapter is, therefore, to identify and examine those processes which could affect sediment supply to streams draining forested areas. Three catchments were studied in Luxembourg, each with a predominantly beech or mixed oak/beech woodland cover, but draining contrasted lithologies. Different sediment-supply processes could be evaluated since their relative importance reflected the particular forest environment of a given catchment.

REGIONAL SETTING OF THE LUXEMBOURG CATCHMENTS STUDIED

1. Geology and Relief

The small Haarts catchment (0.169 km^2) lies southwest of Wiltz (Figure 6.1) on the highly folded, weakly metamorphosed Devonian shales of the Luxembourg Ardennes. This area is a level plateau at an elevation of about 500 m. Steeply incised valleys create a relief amplitude of about 95 m, slopes are convex, steepening to maximum angles of about 35° just above the stream channel.

Gutland, the southern part of Luxembourg, differs in many respects from the Ardennes, with Mesozoic sediments of the Paris basin forming a cuesta landscape. The Deifenbaach catchment (0.55 km^2) is located on the Keuper Marls (Steinmergelkeuper) outcrop, midway between Larochette and Ettelbrück, a relatively low-lying region with a relief of

---

[*] Physical Geography & Soils Laboratory, University of Amsterdam

Forested catchments in Luxembourg

Location	Details	Loss/Yield	Source
Oregon, Alsea river basin	Deer Creek	29.4	Williams, 1964
	Flynn Creek	21.2	
	Needle Creek	15.9	
Colorado, central snow zone	Lexen Creek (lodgepole pine)	2.95	Leaf, 1971
	Deadhorse Creek (spruce fir)	3.6	
Oklahoma	(natural woodland)	2.5	Smith & Stamey 1965
North Carolina	(natural forest)	0.5	
Texas	(protected woodland)	2.5	
Luxembourg	(splash erosion only)	1.0	Imeson & Kwaad, 1976
Queensland	Barron catchment	15.1	Douglas 1967
	Milstream catchment	16.4	

Table 6.1 Examples of sediment yields or soil losses reported from forested catchments or plots, in tonnes/km^2/year

about 45 m. In most of the wooded areas, the gently sloping valley sides end abruptly in sharply incised, meandering stream channels. Possibly artificial improvement to soil drainage have caused this phase of recent erosion (Imeson & Jungerius 1977).

The Honsschlaed catchment (.24 km^2) is located on the scarp slope of the cuesta formed by the Luxembourg Sandstone. Part of the catchment is located on the overlying Arienten Formation, a marl with interbedded quartzite bands.

2. Climate

Luxembourg has a cool and maritime climate with rain in all seasons with no pronounced maxima throughout the year. There are some 200 days with precipitation annually. The average precipitation is about 880 mm/year in the Ardennes in the northern part of the country. In the Gutland the spatial distribution of the average precipitation is very variable. In the Deifenbaach as in the Honsschlaed catchment averages are 730 mm/year. There are few data available on rainfall intensities. Lahr (1964) states that a rainfall intensity of 1 mm/min occurs rarely in the Grand Duchy. The annual number of days with freezing conditions is in the north approximately 100 and in the south 90. The dominant wind direction over the whole of Luxembourg is south-west, and from that direction also the highest average and maximum wind speeds are recorded. Highest mean values are recorded in winter.

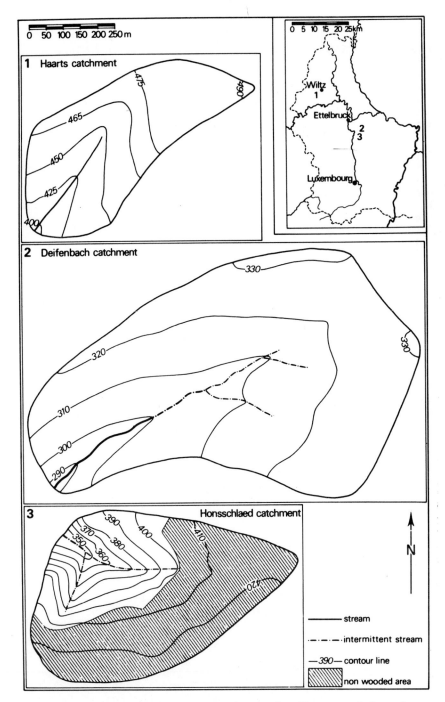

Figure 6.1 Location and relief of the three catchment areas.

3. Soils

On plateau tops in the Haarts catchment, shallow truncated soil profiles are found, but on most of the slopes the soils (dystric Cambisols) and regolith are relatively deep. The soil surface and the subsoil are stony to very stony, with 40-60 per cent being coarser than 10 mm in size. Silt makes up 30-45 per cent and clay generally forms 10-20 per cent of the soil. The texture of the material finer than 2 mm varies little.

Under forest, the soil profiles in the Deifenbaach catchment are characterized by a sharp boundary between a rather silty, permeable A2 horizon and an impermeable clayey subsoil. In the Honsschlaed catchment, the Arieten Formation weathers into material having a frequency distribution with a modal value of less than 50 µm, whereas the modal value of the weathering products of the underlying sandstone is in the 105-210 µm size range. Orthic Podzols have developed on the sandstone slopes with a southerly exposure but elsewhere on the sandstone and the Arieten Formation, chromic or dystric Cambisols are found.

4. Vegetation and land-use

The Haarts catchment is one of the few completely forested drainage basins in the area. Elsewhere, the plateau surface is generally cultivated. Most of the steep valley sides are wooded with oak/beech forest and spruce or fir plantations. There is little vegetation beneath the near-complete canopy, but a thick litter layer covers the forest floor. Similarly, the Deifenbaach catchment is completely wooded with mixed oak-beech forest, with occasional young spruce plantations. In the Honsschlaed catchment, the steep sandstone slopes (20-40 degrees) and the gentler slopes of the Arienten Formation have a mixed vegetation of beech (>75 years old), 20-year old pine and 5-year old spruce woodlands. As in the other catchments, except on the steepest slopes, the forest floor is almost completely litter-covered. Above the woodland the catchment is used as arable farmland with wheat, barley and maize being the principal crops.

5. Pedohydrology and stream flow

Precipitation infiltrates rapidly into forest soils, partly because the litter layer and the stability of soil aggregates protect the soil from compaction and crust formation which may be induced in agricultural soils. Equally, the continuity of the biopores, created by soil fauna and tree roots, is not disturbed. In Luxembourg, the relatively coarse texture of the A horizon of the forest soils, even where these have formed on predominantly argillaceous material is another factor favouring infiltration. In the Honsschlaed catchment, infiltration rates of the A horizons vary between 155 and 7900 mm/hr (van Zon 1978). In the Scheisgrond catchment, adjacent to the Haarts catchment, infiltration rates are 3400-27500 mm/hr in the A horizon and 1500-3300 mm/hr in the B horizons, variations occurring within a few square metres (Kwaad 1977).

Most of the water draining the Haarts slopes moves rapidly through highly permeable zones at a depth of 2-3 m near or at the base of the weathering material, reaching the river a day or so after rainfall (Verstraten 1977). Such drainage discharges at springs or poorly defined seepages directly or almost directly into the river. Since maximum seepage, which might be interpreted as return flow, lags at least a day after precipitation, conditions favouring saturated overland flow are seldom attained. On the Deifenbaach catchment slopes unconcentrated overland flow occurs after heavy rain, wherever soil conditions are favourable. Such flows, however, are largely confined to the numerous micro-topography depressions and seldom reach the river. Despite the impermeable clayey subsoil and the waterlogging of surface soils, the stream is intermittent even in winter. In the highly permeable soils of the Honsschlaed catchment, repeated attempts to detect overland flow or throughflow on the sandstone slopes have been unsuccessful (van Zon 1978). As in the Deifenbaach catchment, the stream only flows after heavy rain.

SOURCES OF SEDIMENT IN FORESTED CATCHMENTS

1. Soil fall

The process of soil fall (Imeson & Jungerius 1977) is clearly displayed in the Deifenbaach catchment, where the undercutting stream or drainage ditches expose soil profiles. Soil falls due to the marked contrast in erodibility of the surface and subsoil horizons, and through the rooting characteristics of the vegetation. Root penetration of the subsoil is impeded by its clayey texture, waterlogging, shrinking and swelling and by some fragipan development. In consequence, a root mat has developed at the base of the A2 horizon. When the subsoil is exposed to the atmosphere, it readily disintegrates into an incoherent mass of angular aggregates a few mm in size. These fall off and the subsoil exposure retreats beneath the root mat. This process is important for sediment supply since 35 per cent of the channel bank is undercut in this way, contributing some 10 per cent to the suspended sediment output of 210 kg in 1977. With recessions of a metre or more, depending on the strength of the root mat, the surface soil and root mat collapses, covering the previously eroding slopes and thus arresting erosion at the point for some time. However, subsoil material may occasionally be exposed again if the root mat is broken at the point of collapse, and recession may recommence at a higher level.

Erosion pin measurements suggest that the rate of recession and sediment supply are highly seasonal. The timing of the sediment-supply maximum depends on whether material falls directly into the river channel or accumulates on the flood plain or lower terrace. In the first case, most material is released from the backwall during the early winter and spring. Also, the supply of material from the A2 horizon and overhanging root mat is important during periods of soil moisture surplus. This material tends to be transported directly by the stream. Unless the discharge is

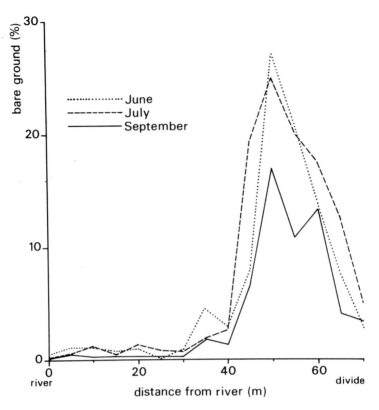

Figure 6.2 Bare ground in the Deifenbaach catchment in June, July and September, 1975

relatively high, however, aggregates from the backwall will accumulate beneath the water surface where they appear to be relatively water-stable.

2. Exposed mineral soil on slopes

Beneath the forest canopy in the Luxembourg forests, bare mineral soil occurs, first, on steep slopes where litter either fails to accumulate or is easily displaced by wind, and secondly loose, bare soil may occur wherever the litter layer is disturbed by animal activity. Generally such areas occupy 0.01-5 per cent of the forest floor in the Ardennes but can approach 80-90 per cent in the Gutland.

Burrowing rodents frequently bring up structureless materials from the B and C horizons to the ground surface. The activity of burrowing rodents was measured on plots set out in the Deifenbaach catchment in the summer and autumn of 1975. The plots, 5 m wide, were set out one beneath the other from the catchment divide downslope to the recent channel incision. The proportion of bare ground was recorded (Figure 6.2) and attributed to the burrowing of moles and voles and in some cases to deer or worms. On

Forested catchments in Luxembourg

other slopes not investigated in detail worms are found during wet periods in the summer and autumn to cover as much as 80-90 per cent of the ground surface with their casts. On the slope studied, however, the precise origin of the bare mineral soil exposures was difficult to determine as they were soon eroded, often within a few weeks. This transect passed over a varied micro-topography, with one large and another small depression in which soil conditions were moist and animal activity particularly prominent. Such activity was also greatest in the damper parts of the Haarts catchment where this damper zone tends to occur along the river. In the Haarts catchment, burrowing reaches a peak in early summer (Fig 6.3). However, the total of bare ground reflects more the yearly cycle of litterfall and decay rather than periodicity in animal burrowing. Thus, in the autumn in the Haarts catchment, animal mounds are covered with litter. In fact, the breakdown or removal of this litter results in old mounds being re-exhumed in the late spring and summer so that areas of exposed soil are at a maximum before the period of leaf fall (Imeson & Kwaad 1976).

TRANSLOCATION OF SEDIMENT DOWNSLOPE AND INTO STREAM CHANNELS

1. **Splash erosion**

In the Haarts catchment, the supply of material to the river by splash erosion was estimated at approximately 190 kg/yr in 1974. About 95 per cent of the sites supplying splashed material appeared to be the result of animal activity and 50 per cent of these sites occurred within 25 m of the river on the steep lower slopes. When the data from plots in the Haarts and Diefenbaach catchments are compared it seems that the latter has about 50 per cent more bare ground. This larger area and the greater erodibility of the soil, there being no stones to protect the excavated material, make splash perhaps a more important process on the slopes of the Deifenbach catchment, and storm water translocates high concentrations of dispersed clay. However, splash contributions to the river are probably less since most mounds are situated at some distance from the river channel and because the slopes are less steep. Close to the stream banks, the aggregates produced by soil-fall are susceptible to splash erosion, especially in accumulations at the foot of recessions above flood plains or terraces. Upon exposure to splash, a crusted surface soon develops. In the Honsschlaed catchment, where there is little evidence of burrowing by rodents, amounts of splashed material vary greatly, despite apparently similar ground-surface conditions. Nonetheless, an exponential increase in net amount of material transported with increasing slope angle was detected (Figure 6.4). On a 30 degree slope, twenty times as much material was transported in a year from ground kept bare as from a slope surface with a litter cover of 70-100 per cent on the same parent material (200 gm as compared to 10 gm). Loss from the 30 degree bare slope plots was approximately 70 times that from a completely litter-covered slope of 2

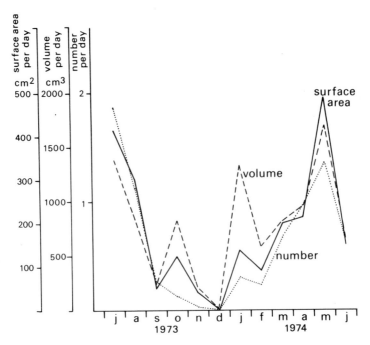

Figure 6.3 Variation in burrowing activity on plots in the Haarts catchment

degrees where only some 3 gm splashed downslope (van Zon 1978).

The translocation of sediment by splash erosion is highly seasonal. In the Haarts catchment, maximum rainfall efficiency occurs during the summer, indicating that the canopy increases the erosivity of the rain (Kwaad 1977). Thus, most material is supplied from the lower slopes to the river in the late summer and autumn when splash erosion is most effective (Fig 6.5). Similarly, the key factor in the Honsschlaed catchment, the area of bare ground, shows a yearly cycle which peaks in early autumn. In the Deifenbaach catchment, more bank-fall material accumulates in spring and summer than is lost by splash erosion. In winter, the converse is the case, but seasonal differences are not large.

2. Leaf transport

Leaf transport provides a mechanism whereby sediment splashed from soil exposures located anywhere beneath a forest may be relatively rapidly supplied to the drainage courses. In the Honsschlaed catchment, heavy rain (>10 mm/h) occasionally displaces litter which becomes re-arranged in concertina-like forms. Such movements often occur downslope from beech trees, and presumably reflect the occurrence of stemflow on their smooth trunks. Due to splash erosion, such leaves may have mineral particles adhering to their surfaces. Two main types of event can

Forested catchments in Luxembourg

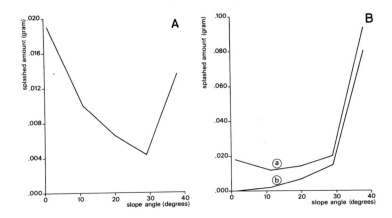

Figure 6.4   A   Amount of material splashed in an upslope direction as a function of slope angle

B a) amount of material splashed in a downslope direction as a function of slope angle
   b) net amount of downslope transported material by splash as a function of slope angle

initiate the movement of such leaves lying loose on the forest floor (van Zon 1978). First, leaves may be dislodged by occasional, extreme events, such as localized overland flow, a severe gale, or trampling by a herd of deer. Such events occur irregularly in time and may not be concentrated at specific locations. Secondly, minor but frequent movements of a more continuous nature are initiated by wind or rain. Monthly collections of moving litter and the sediment adhering to it revealed that leaf transport varies with type of leaf, slope angle, and with various microclimatic factors (Figs 6.6 and 6.7). The thickness of the transportable leaf cover on the slopes is important. This is a very variable quantity, depending on antecedent conditions. Inevitably, it is greatest during the autumn period of leaf fall when the leaves are very susceptible to transport. They do not, however, necessarily transport much sediment since they mainly remain close to areas of bare soil only temporarily. During a measurement period in 1976, in autumn 0.10 gm sediment was translocated per 10 grams of transported leaves, compared to 1.5 gm during the late summer period. The amount of sediment transported by a given weight of leaves is dependent on the amount of bare soil from which the sediment is supplied (van Zon 1978). When the variation attributable to precipitation is eliminated by partial correlations, the amount of bare soil emerges as a more important variable than the erosivity of the rainfall. The transport of leaves tends to decrease after autumn, but the seasonality is not pronounced. The amount of sediment transported by leaves shows a late-summer - early-autumn peak, reflecting the changes in areas of bare ground and in the erosivity of the rainfall.

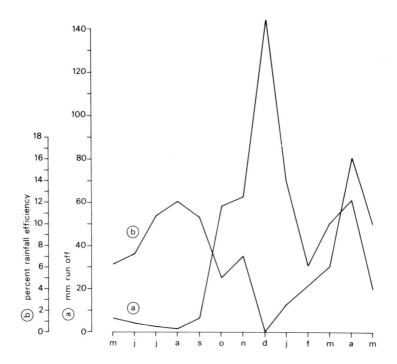

Figure 6.5 Rainfall efficiency and runoff in the Haarts catchment

3. Other processes of sediment translocation

The two sediment-supply processes described above, although their action is prominently displayed in the three catchments under discussion are, however, not the only ones supplying sediment to streams in forested catchments, even within an area as small as Luxembourg. For instance, in the Haarts catchment, material excavated by animals and subsequently moved by the effect of gravity alone was estimated as about 90 kg/yr, almost half the quantity supplied by splash.

EVACUATION OF SEDIMENT
FROM THE STREAM CHANNELS

Clearly, the yearly cycle of sediment entering the stream channel may be out of phase with the discharge and its pronounced winter maximum (Figs 6.3 & 6.5). During individual summer storms in the Haarts catchment, sediment pulses and flood waves are only approximately in phase with splash erosion on the large area of bare soil along the river channel which produces high sediment concentrations (1500-2000 mg/litre) (Imeson 1977). Furthermore, relatively insignificant rainfall events (2-5 mm), insufficient to raise stream discharge, may nonetheless produce almost comparable high sediment concentrations. At other times, sediment pulses may be produced by burrowing animals

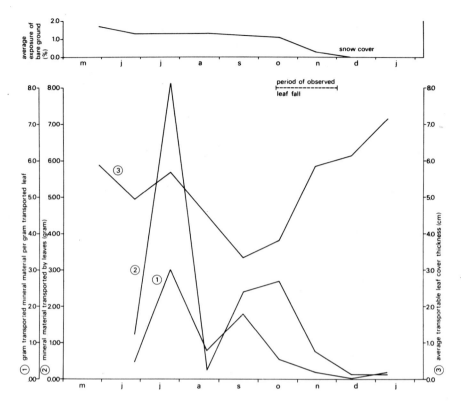

Figure 6.6 Average exposure of bare ground, average transportable leaf cover thickness, mineral material transported by leaves and gram transported material per gram transported leaf, over the period May 1976 - January 1977 on a north-exposed measurement site under 70-year old beech woodland

excavating material directly into the stream or by larger animals drinking or trampling in the stream.

During the winter in the Haarts catchment, peak discharges occur about 24 hours after rainfall, with the water drainage from the weathering materials and deposits on the catchment slopes. In autumn and spring, double hydrograph peaks may occur as the water from the weathering zone reaches the stream about 12-14 hr after the initial peak produced by channel precipitation. This water could be either groundwater or interflow, but whilst creating high discharges, sediment concentrations are very low (<20 mg/litre). Occasionally, flows are sufficient to transport the gravel which accumulates in the river channel, supplied from the slope deposits of the channel banks and from the lengthy exposures of shale in the channel itself. Splash erosion can cause particles larger than 10 mm in size to roll into the river, and gravel is also supplied by animal activity on the banks and by

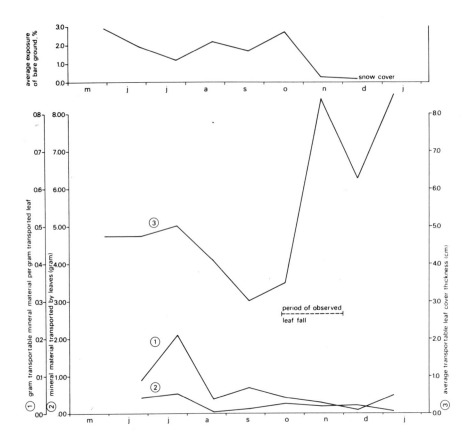

Figure 6.7 Average exposure of the ground, average transportable leaf cover thickness, mineral material transported by leaves and gram transported material per gram transported leaf over the period May 1976 - January 1977 on a south-exposed measurement site under 70-year old beech woodland

movements initiated by needle-ice growth. Usually most of the gravel smaller than 5 cm in size is removed from the steep 7-8 degree channel by the peak discharges which occur once or twice a year. At such times, sediment transport and discharge are related, even though the material may have been supplied to the river much earlier.

Since the Honsschlaed catchment stream flows only a few times each year, leaves and splashed material also accumulate in the stream channel. Both leaves and sediment are transported whenever the stream flows, the leaves being especially easily transported by flowing water. The depth of leaf accumulation was measured at monthly intervals along the stream channel and average values plotted (Fig 6.8), together with discharge, precipitation and leaf transport data. Clearly the amount of leaf transport reflects the

period	25/5-25/6	25/6-24/7	24/7-20/8	20/8-20/9	20/9-15/10	15/10-13/11	13/11-10/12
transported leaves (gm)	4.79	162.0	4.32	5.43	151.22	286.32	?
transported mineral material by leaves (gm)	.29	51.82	.39	.35	18.59	14.69	
discharge (m^3)	.00	.84	.00	.00	1.13	9.62	4809.56
precipitation (mm)	36.6	98.7	9.3	49.8	59.3	68.5	105.4

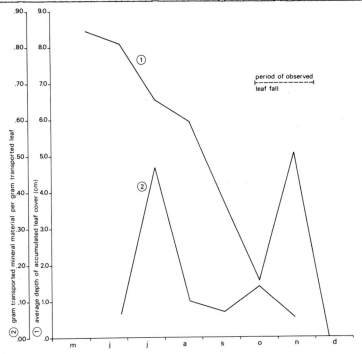

Table 6.2 and Figure 6.8 Average depth of accumulated leaf cover and grams transported mineral material per gram transported leaf in the river bed over the period May 1976 - December 1976

volume of discharge (Table 6.2). In particular, an exceptional discharge in December 1976 removed all the leaves which had accumulated during the preceding 8 months. The discharge of sediment reflects the occurrence of discharge events which occur much more frequently in the winter.

CONCLUSIONS

In general, forests in humid temperate regions have permeable A horizons and a near-continuous cover of litter and ground vegetation which restrict the occurrence of both overland flow and splash erosion, explaining the low erosion rates. In detail, however, soil and ground cover conditions beneath forest may be complex and both mechanisms can operate locally. Obviously, sediment transport by processes such as animal burrowing or leaf movement are dependent upon various characteristics of the forest ecosystem, since the precise nature of the vegetation influences raindrop fall and interception, and reflects soil and drainage conditions. When the timing of the sediment supply to the stream channels of the three Luxembourg catchments is considered, it emerges that this is often out of phase with sediment transport. This occurs because sediment is supplied by processes which do not necessarily coincide with runoff-generating events. One consequence of this is an inexact relationship between water discharges and sediment concentrations and even, in some instances, with sediment discharge. In general the summer half of the year may be considered as the period of greatest erosion and sediment supply to the river, and the winter half of the year, the period of sediment evacuation. Finally, in all the catchments, the areas of bare soil play a most important part in sediment supply. The generation and persistence of such mineral soil exposures under various forest conditions is a critical subject for further investigation.

ACKNOWLEDGEMENTS

We are indebted to Professor P. D. Jungerius and Dr. F. J. P. M. Kwaad for reading the text, and Dr. A. F. Pitty for revising and editing the manuscript. We also wish to thank Mr. C. Snabilié and Mr. van Geel for drawing the illustrations.

REFERENCES

Douglas, I., 1967. Natural and man-made erosion in the humid tropics of Australia, Malaysia and Singapore. *Publication of the International Association of Scientific Hydrology,* 75, 17-29

Hudson, N., 1973. *Soil conservation.* (Batsford, London)

Imeson, A.C., 1977. Splash erosion, animal activity and sediment supply in a small forested Luxembourg catchment. *Earth Surface Processes,* 2, 153-160

Imeson, A.C. & Jungerius, P.D., 1977. The widening of valley incisions by soil fall in a forested Keuper area, Luxembourg. *Earth Surface Processes,* 2, 141-152

Imeson, A.C. & Kwaad, F.J.P.M., 1976. Some effects of burrowing animals on slope processes in the Luxembourg Ardennes. Part II: the erosion of animal mounds by splash under forest. *Geografiska Annaler,* 58 (Series A,4), 317-328

Kwaad, F.J.P.M., 1977. Measurement of rainsplash erosion and the formation of colluvium beneath deciduous woodland in the Luxembourg Ardennes. *Earth Surface Processes,* 2, 161-173

Lahr, E., 1950/1964. *Temps et climat au Grand-Duché de Luxembourg* (Publication du Ministère de l'Agriculture Luxembourg)

Leaf, C.F., 1970. Sediment yields from the central Colorado snow zone. *Journal of the Hydraulics Division, American Society of Civil Engineers,* 96(HY1), 87-93

Lull, H.W., 1971. Runoff from forest lands. In: *Man's impact on terrestrial and oceanic ecosystems.* ed. W. H. Matthews, F. E. Smith & E. D. Goldberg (MIT Press, Cambridge, Massachusetts), 240-251

Smith, R.M. & Stamey, W.L., 1965. Determining the range of tolerable erosion. *Soil Science,* 100, 414-424

Toy, T.J. (ed), 1977. *Erosion: research techniques, erodibility and sediment delivery.* (Geo Books, Norwich)

Verstraten, J.M., 1977. Chemical erosion in a forested watershed in the Oesling, Luxembourg. *Earth Surface Processes,* 2, 175-184

Williams, R.C., 1964. Sedimentation in three forested drainage basins in the Alsea River basin, Oregon. *United States Geological Survey, Geological Survey Circular,* 490

Zon, H.J.M van, 1978. Litter transport as a geomorphic process. *Publicaties van het Fysisch-Geografisch en Bodemhundig Laboratorium van de Universiteit van Amsterdam,* 24

CHAPTER 7 RELATIONSHIPS BETWEEN PLASTICITY NATURAL MOISTURE CONDITIONS AND SURFACE STABILITY OF SOME SLOPE SOILS NEAR HELMSLEY NORTH YORKSHIRE

Roger Cooper[*]

INTRODUCTION

Movements of individual soil particles towards stream channels are measurable over a year, and are even detectable during shorter intervals. Apart from relatively continuous creep and wash processes, a major source of sediment supplied to stream channels is the abrupt slide of a large volume of material *en masse*. As such slides may be witnessed only at intervals of decades or even centuries, the mechanisms involved evade deliberate measurements. Frequently the details of processes may be reconstructed where a slide has just chanced to occur. However, a second problem is that, with landslides taking place as widely separated, individual events, the local conditions tend to be varied, and therefore generalization is difficult. An alternative approach to the study of soil stability in natural conditions may be made via the monitoring of the varying natural inputs of moisture, the critical factor in inducing degrees of instability. An area prone to landslipping is selected and soil moisture monitoring carried out in addition to measurement of the static properties which account for the stability of soils. By recording the degree to which natural moisture conditions increase episodically and approach critical limits of soil stability, some impression of the general susceptibility for mass transfer of materials downslope can be gained. The definition of a study area in which several key factors, such as geology and slope angle are similar, and one of sufficiently small scale for climate and vegetation to be the same, limits the problem of the variability of conditions under which slides are actually observed. Sampling at several points within such an area provides a sound basis for generalization. The present study is concerned with the plasticity and natural moisture conditions in soils at 24 sites on slopes in the Hambleton Hills, near Helmsley in North Yorkshire. Relationships between indices of these attributes are examined, both directly and indirectly, and in relation to particle-size distribution and the soil moisture fluctuations induced by seasonal variations in precipitation. The results are examined in the context of mass-movement studies in the area.

[*] Department of Geography, University of Keele (now at Department of Arts & Social Sciences, Dorset Institute of Higher Education)

Figure 7.1 General view of valley slopes and plateau top in the Hambleton Hills. Ryedale occupies the foreground. In the middle ground the wooded slope is the site of the Ashberry Windypits and slope profile B (Figure 7.7). The plateau top of the Hambleton Hills makes up the background.

THE STUDY AREA

1. Geology, relief and drainage

The Hambleton Hills form the western edge of the uplands of north-east Yorkshire. 30 km to the north of York, they rise to 374 m OD in an easterly-tilted plateau. The plateau is deeply dissected by east-flowing tributaries of the River Rye (Fig 7.1), the valley of which forms the eastern boundary of the area (Fig 7.2). The hills are made up of well-jointed Corallian rocks of the Upper Jurassic, which consist of alternating bands of spicular sandstone and oolitic limestone (Fox-Strangways 1892), and which are near-horizontal in disposition in this area. The underlying Oxford Clay (silty clay) and Kellaways Rock (sandstone) outcrop at the foot of the slopes, and on the valley floors where they are often obscured by alluvium. There is little surface drainage on the plateau top, which is formed mainly in oolite. Springs are thrown out on the valley sides of the Rye and its tributaries, and on the western scarp face of the hills, by impermeable bands in the Corallian and at points in a broad stratigraphic transition zone (decreasing silt/clay content upward) from the Oxford Clay to the Corallian. These geological and associated geohydrological conditions render the valley slopes in the area particularly susceptible to mass movements of soil and rock materials.

Slope soils near Helmsley, N. Yorkshire

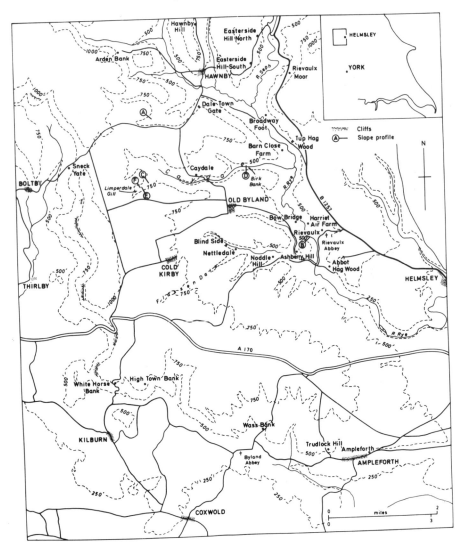

Figure 7.2 The Hambleton Hills, showing sites and localities mentioned in the text

2. Climate

A range of climatic observations has been recorded at Ampleforth (SE598788), 5 km south of Helmsley. Long-term average monthly mean temperatures (1931-60) vary between 2.7°C (January) and 15.5°C (August), with an annual mean of 8.8°C (Meteorological Office 1963). Long-term average precipitation totals (1941-70) at the several rainfall recording stations in and around the Hambleton Hills vary between about 650 mm and 800 mm, according to altitude, with little variation throughout the year, and a maximum usually in August (Yorkshire Water Authority *personal communication* 1975). Snow lies for 30 to 45 days per year. Average

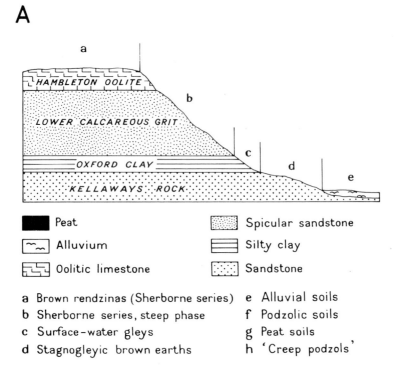

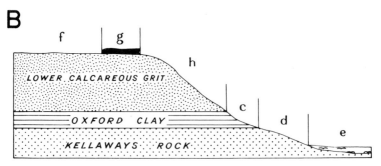

Figure 7.3 Influence of bedrock lithology and steepness of slope on soil types in the Hambleton Hills.
a) with oolite caprock
b) without oolite caprock

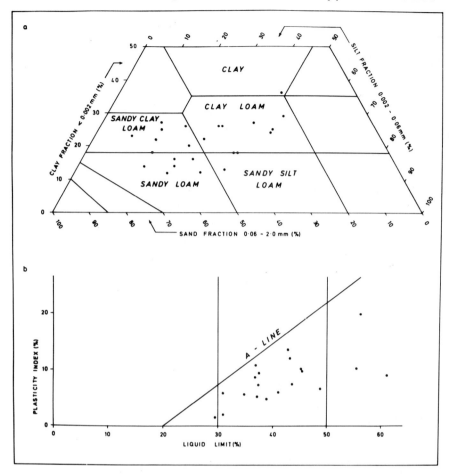

Figure 7.4 a) Proportions of clay, silt and sand in the fine earth (<2mm) fraction of soils sampled, according to the classification described by Hodgson (1974).
b) Plasticity chart.

annual evapotranspiration varies from 430 mm on high land to 480 mm in lower areas. In an average year it will exceed precipitation during the summer months (Bendelow & Carroll 1976).

3. Soils

The soils on the extensive areas of oolitic limestone are almost entirely rendzinas of the Sherborne series (Bendelow & Carroll *in press*). These lithomorphic soils have typical A/C profiles, with a dark brown, loamy or stony A_p horizon about 20 cm thick, overlying the weathered bedrock. There is some interstitial dark brown

Slope soils near Helmsley, N. Yorkshire

clay loam, and locally there are admixtures of loess (Catt et al 1974). Elsewhere on the plateau, podzols predominate where sandstone outcrops, and some peats have developed in the highest areas.

Soils in the tributary valleys of the Rye, such as Gowerdale, Caydale and Nettledale, are distinguished as a 'steep' phase of the Sherborne series by the Soil Survey. In detail, soils of the steep slopes are much affected by soil creep. Although the bedrock may be spicular sandstone, the soil is often dominated by downward-creeping fragments of oolitic limestone from an outcrop at the shoulder of the valley sides, and brown rendzinas develop (Anderson 1958). Such soils are generally free-draining, and gritty clay loams in texture. 'Creep podzols' develop where the limestone outcrop is not present. These soils are excessively drained and silty fine sand in texture. Soil creep tends to minimise A horizon differentiation. At the outcrop of Oxford Clay, the mixture of clay with the colluvium, reduced runoff rates, and downslope seepage cause gleying. Gleyed brown earths are found on the Kellaways Rock, and many valley floors are covered by alluvium (Fig 7.3).

4. Vegetation and land-use

The pattern of moorland plant communities in the Hambleton Hills is related closely to the underlying geology and soils. *Festuca-Agrostis* grassland is found on the predominant oolite, while the outcrops of spicular sandstone in the higher northern areas are colonised by Callunetum, associated in place with *Nardus stricta*. In the highest areas there are areas of blanket peat. Many valley sides are now covered with coniferous plantations. Undergrazing of slope pastures leads to colonization by *Deschampsia flexuosa* and *Pteridium aquilinum* (Anderson 1958). The particular types of vegetation found are largely a function of land-use. In addition to tree planting, Callunetum is maintained by periodic burning of grouse moors (Imeson 1971), while *Festuca-Agrostis* grassland is maintained by grazing.

SOIL PLASTICITY AND FIELD MOISTURE CONTENT

1. Definition of the Atterberg limits

The consistency of a soil is the manifestation of the forces of cohesion and adhesion acting within it at various moisture contents (Pitty 1979). On drying, a soil may pass through several stages of consistency, which can be described as the liquid, plastic, semi-plastic and solid stages. Changes in consistency do not take place sharply at precise moisture content levels, and therefore arbitrary moisture content limits proposed by Atterberg (1911) have been adopted. The lower limit of plastic consistency is known as the *plastic limit* (PL) and the upper limit as the *liquid limit* (LL). Plasticity itself is measured by the *plasticity index* (PI), which is simply the difference between the liquid and plastic limits. The plastic limit is thought to be the minimum moisture content at which the adsorbed films of water around clay particles in the soil are sufficiently thick to facilitate intra-particle sliding when an external pressure greater than the tension of the water films is applied (Baver et al 1972). The liquid limit is thought to be the minimum moisture content at which the thickness of the films is sufficiently large to reduce the cohesion between the particles to the extent that the soil mass will flow under external pressure.

The liquid and plastic limits are expressed in terms of $\underline{m}$, the soil moisture content (per cent dry weight). Thus, the *liquidity index* (LI) is a means of expressing $\underline{m}$ in terms of the plasticity index:

$$LI = (\underline{m} - PL)/PI$$

At the liquid limit, a soil will have a liquidity index of 1.0 and at the plastic limit it will have a liquidity index of 0.0. Negative values indicate soil moisture contents below the plastic limit.

2. Factors affecting plasticity

Atterberg recognised that the proportion of clay in a soil sample influences its plasticity. Seed et al (1964) showed that an increase in the proportion of clay causes both plastic and liquid limits to rise, but the liquid limit rises the further, so that the plasticity index also increases. The *activity* of a clay was defined by Skempton (1953a, b) as the ratio of the plasticity index to the proportion of clay-sized particles:

$$\text{activity} = PI / \text{per cent clay}$$

The clay fraction in this case is the amount of clay expressed as a percentage of the fraction which is finer than 0.425 mm. Other factors which may complicate relationships between soil texture and plasticity, such as particle shape, varying proportions of clay minerals or exchangeable cations, are essentially constant in a study which focusses attention on a small, geologically homogeneous area.

3. Field moisture content

The moisture content of a soil, as it occurs in its natural setting, can be termed the 'field moisture content', as opposed to the hypothetical 'field capacity', which is the moisture content after an excess of artificially added water has drained away under the influence of gravity. In contrast, 'field moisture content' includes moisture contents perhaps measured before all excess water has drained away, or, alternatively, moisture contents when soils have dried out below the moisture level of field capacity. The critical characteristic of the field moisture content of a surface soil is that it encompasses natural fluctuations, and measurements spanning seasonal ranges of weather conditions are needed to estimate its mean value.

SAMPLING AND ANALYSIS

The 24 sites chosen for study (Fig 7.2) were located on slopes at positions where the Corallian grades downslope into the Oxford Clay outcrop. Samples at each site were taken from a constant depth below the soil surface. At most sites, samples were taken on 18 occasions - at 2-4 week intervals in the period March 1974 to April 1975. On each occasion field moisture content was taken as the mean of gravimetric determinations on three replicate samples obtained from each site. Liquid limit was determined using the Casagrande apparatus, and it was possible to determine the plastic limit (British Standards Institution 1967) at all but four sites, as the mean of three replicate samples. The remoulding of soil samples for the determination of the Atterberg limits is not believed to introduce critical departures from field conditions, as the soils at their slope sites are subject to creep, mass-wasting and other natural disturbances. The particle size distributions of the soils were determined by wet sieving (coarse fraction) and the hydrometer method (fine fraction), and summarized by the descriptive statistics outlined by Folk & Ward (1957). Samples were not treated to remove organic material or carbonates, as this would have eliminated attributes considered to be important in the field behaviour of the soils.

RESULTS

1. Mean values and main relationships

Statistics descriptive of five characteristics of the sampling sites and 16 properties of the soils are listed (Table 7.1). In particle size, the soils are poorly sorted loams, with varying admixtures of sand, silt and clay (Fig 7.4A). Most of the soil properties tabulated vary considerably between sites as described by the coefficient of variation, CV. Nonetheless, all soils fall within the same small area of the plasticity chart (Fig 7.4B). The mean field moisture content is 16.54 per cent below the mean liquid limit, and also below the mean of the plastic limits at those sites where the latter index could be measured.

Relationships between the soil properties and site characteristics are summarized (Table 7.2). The fundamental control of the proportion of clay on soil plasticity (PI) is clearly demonstrated. Conversely, the coarser the soil texture, as expressed by the per cent sand, the lower the liquid limit. Plasticity and activity decrease with the distance of the sampling sites up the slope. The liquid and plastic limits correlate positively with maximum field moisture content, and with mean field moisture content and its standard deviation. These relationships suggest that those soil characteristics which account for high liquid and plastic limits tend also to increase the chances of relatively high mean and maximum field moisture contents being achieved, and hence to accentuate contrasts in the moisture content of the soil between wet and dry periods.

Property	n	mean	minimum	maximum	sd	CV(%)
LL (%)	24	38.85	10.5	61.0	10.62	27.3
PL (%)	20	33.25	25	52	6.75	20.3
PI (%)	20	8.37	1.5	20.0	4.17	49.8
activity	20	0.34	0.09	0.69	0.15	43.9
mean $\underline{m}$ (%)	24	22.33	10.57	37.40	5.80	25.9
s.d $\underline{m}$ (%)	24	3.92	1.41	9.38	1.97	50.3
max. $\underline{m}$ (%)	24	29.42	13.34	51.63	7.21	24.5
min. $\underline{m}$ (%)	24	15.37	5.48	27.45	5.26	34.2
mean LI	20	-1.58	-4.06	-0.21	1.19	75.4
clay (%)	24	21.21	12	36	6.20	29.2
silt (%)	24	30.38	10	48	11.25	37.0
sand (%)	24	48.42	20	68	14.07	29.1
mean size (ϕ)	24	5.54	3.40	7.43	0.83	15.0
sorting (ϕ)	24	3.74	3.10	4.53	0.32	8.6
psd* skewness	24	0.46	0.14	0.85	0.17	36.2
psd* kurtosis (normalised)	24	0.53	0.43	0.73	0.07	13.2
sampling depth (m)	24	1.08	0.3	2.1	0.53	49.1
slope angle	24	21°	7°	44°	10.35°	49.3
altitude (m OD)	24	149.5	76	213	44.87	30.0
height above thalweg (m)	24	38.0	2	110	32.63	85.9
length of slope (m)	24	196.9	52	423	97.97	49.8

* psd = particle size distribution, psi units

Table 7.1 Variation in measured soil properties and site characteristics

2. Seasonal changes of the liquidity index in relation to the liquid and plastic limits

The fluctuating moisture contents of those soils of which it was possible to determine the plastic limits are shown in Fig 7.5 as liquidity indices. They are grouped for convenience according to amplitude of fluctuation, which depends on, to a great extent, plasticity index. The latter varies widely between sites, so it is not meaningful to compare amplitudes of fluctuation. However, the graphs illustrate clearly the movements of field moisture content towards and away from the critical levels represented by the liquid and plastic limits. In general, those sites with greater variability (Table 7.1) showed a fairly regular seasonal trend, with 15 sites reaching a maximum in December to March, and 15 reaching a minimum in June or July. The

	LL	PL	PI	activity	mean m̄	s.d. m̄	max. m̄	mean LI	height above thalweg	clay	silt	sand	mean size	psd† skewness	psd† kurtosis
LL	–														
PL	.88*	–													
PI	.63*	.19	–												
activity	.73*	.41	.81*	–											
mean m̄	.65*	.60*	.08	.45*	–										
s.d. m̄	.57*	.53*	.18	.27	.35	–									
max. m̄	.77*	.67*	.09	.40	.86*	.63*	–								
mean LI	.22	–.14	.65*	.64*	.28	–.21	.15	–							
height above thalweg	–.27	–.25	–.48*	–.51*	.07	–.34	–.03	–.26	–						
clay	.43	–.12	.66	.17	–.09	.11	.14	.45	–.33	–					
silt	.51	.56	.18	.17	.42	.21	.38	.09	–.29	.24	–				
sand	–.59	–.42	–.43	–.21	–.29	–.21	–.37	–.26	.38	–.63	–.90	–			
mean size	–.57*	–.02	.64*	.23	.18	.24	.41*	.50*	–.31	.91	.41	–.73	–		
psd† skewness	–.24	–.52*	–.19	–.14	–.06	.03	.08	.09	–.10	.01	–.71	.56	.03	–	
psd† kurtosis	–.38	.15	–.64*	–.32	–.02	–.24	–.21	–.54*	.44*	–.71	–.24	.50	–.64*	–.09	–

†particle size distribution, phi units

*significant at the 5 per cent level (t-test). Statistical significance is not assigned to correlations with the clay, silt and sand proportions of the soils, which are expressed as variables of constant sum. Attributes included in Table 7.1 but omitted from Table 7.2 show no correlations significant at the 5 per cent level with LL, PL, PI or activity. A further site characteristic, aspect, was examined by the use of the chi-square test, but was found not to have a statistically significant influence. PL and s.d m̄ were found to be distributed with skewness and kurtosis significant at the 5 per cent level (t-test), and were normalized by transformation to logarithms to base 10.

Table 7.2 Correlation matrix of selected soil attributes

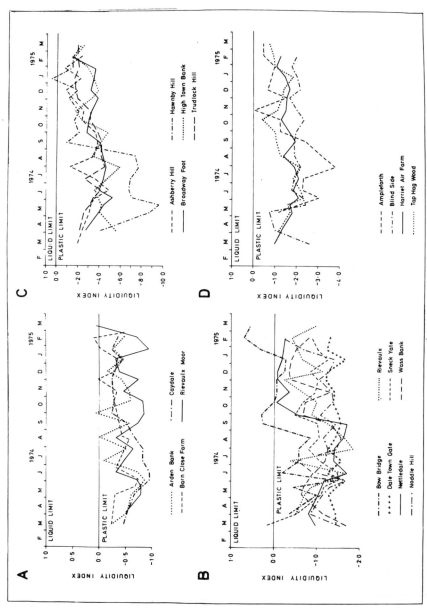

Fig 7.5 Temporal fluctuations of liquidity index in relation to the liquid and plastic limits

Slope soils near Helmsley, N. Yorkshire

	J	F	M	A	M	J	J	A	S	O	N	D
A		84	59	23	39	47	101	61	106	168	77	50
B	76	24	79									
C	82	68	54	59	63	68	77	103	75	73	94	76

A - 1974;   B - 1975;   C - Long-term averages 1916-1950
(Source: Yorkshire Water Authority *personal communication* 1975)

Table 7.3   Monthly rainfall totals at Allen's Garage, Hawnby (mm)

effect of individual monthly rainfall totals (Table 7.3) is evident at some sites. For example, the high rainfall in October produced an immediate rise at the Nettledale site (Fig 7.5B), and the maximum recorded liquidity index (and hence field moisture content) at the Ampleforth site (Fig 7.5D) was reached early in November. Subsequently, despite much reduced monthly rainfall totals, the liquidity indices at these sites remained high, but still negative, until the following March. Conversely, the low rainfall of April and May, coupled with increasingly warm weather, resulted in the low values measured at most sites early in June. Response to short periods of rain was detected at some sites, although long runs of rain-days are necessary to produce a sustained effect on the pattern of field moisture content. For example, the effect of a very wet period during the first four days of July, when 43.6 mm of precipitation fell, was detected at the Trudlock Hill (Fig 7.5C) Arden Bank (Fig 7.5A) and Rievaulx (Fig 7.5B) sites.

The potentially critical circumstance of field moisture content exceeding the plastic limit, giving a positive liquidity index, was observed at six sites for short periods. The liquidity index rose above 0.0 in September, at the Arden Bank (Fig 7.5A) and Bow Bridge (Fig 7.5B) sites. The Bow Bridge site was similarly affected in the following February and March, as were the Barn Close Farm (Fig 7.5A) and Rievaulx Moor (Fig 7.5B) sites. The Hawnby Hill site (Fig 7.5C) reached a condition of potential instability in January, and the Sneck Yate (Fig 7.5B) site was in this condition when first sampled in March. However, plastic flow was not observed at any of the actual field sites on these occasions, possibly due to the binding effect of the vegetation cover.

PROCESSES OF MASS WASTING ON THE SLOPES

Mass movements on slopes in the Hambleton Hills take a variety of forms. In Flassen Dale (Fig 7.2) there is a lateral series of rotational slumps, possibly triggered by concentration of runoff during early stages of reafforestation. At the crest of the western scarp of the hills are a series of cliffs in the Corallian (Fig 7.2) with many joint-bounded blocks, tilted downslope and liable to toppling failure. Toppled blocks litter the screes below the cliffs. However, perhaps the most striking mass movements in the area are apparent large-scale 'block glides' (Varnes 1958) involving entire sections of hillside. Evidence is provided

Figure 7.6

Exploring an underground fissure in the Hambleton Hills, the Ashberry Windypits (Fig 7.1 shows the situation and Fig 7.7B the survey of this fissure)

by the occurrence of widened vertical joints in the Corallian (Fig 7.6) on or near to slopes, and known locally as 'windypits' (Fitton & Mitchell 1950). These can be entered and explored (Cooper 1978), and many have been surveyed (Cooper et al 1976). They seem to represent an early stage in a process whereby large blocks become separated from the parent hillside, move and tilt valleywards, and break up to form screes and oversteepened vegetated slopes. That such a process is initiated by water, and therefore essentially an integral part of fluvial processes, is well-attested (Leopold et al 1964 p 340). Evidence of the later stages is found in the form of ridge-and-trough features (Briggs & Courtney 1972) on the slopes. At many locations in the Hambleton Hills, these have been identified by the Ordnance Survey as disused quarries. However, quarrying in the area leaves distinctive features (Cooper 1977) which are absent from the ridges and troughs. Ridges downslope from troughs are often found to be dipping with a downslope component, and it is unlikely that this could be caused by quarrying. The stages of the process of mass movement may be recognised from the resultant slope forms, which have been recorded on slope profiles measured over 1.5 m base-lengths using a slope pantometer (Pitty 1968). Many stages appear to be represented at different locations in the Hambleton Hills, and selected profiles may be arranged tentatively as a time-sequence, after the manner of Savigear (1952) and Palmer (1956).

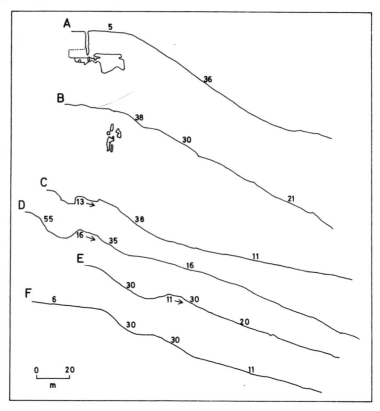

Figure 7.7   Profiles of selected slopes in the Hambleton Hills
Above - examples of slopes with surveyed widened fissures beneath the ground surface, A. Noddle End Windypit; B. Ashberry Windypits (cave surveys from Cooper et al 1976).
Below - a sequence of slope profiles in a tentative time-sequence, C. East side of Limperdale Gill; D. Birk Bank; E & F. West side of Limperdale Gill

The scale of the process in relation to that of the valley-side slopes can be judged from Fig 7.7. It is clear that the fissures separate large volumes of rock from the parent hills. Initially, valleyward movement of rock masses with widening of fissures causes little disturbance of the essentially rectilinear form of the slopes. This is shown in Profile A, over Noddle End Windypit (Fig 7.7A). However, profile B, over the Ashberry Windypits, shows some disturbance of slope form where the smoothing effects of surface movements have been unable to keep pace with the amount of block movement (Fig 7.5B). Profile C (Fig 7.7C) shows a marked trough due to further widening of a fissure, with ridges immediately downslope. The slope below the ridges is steepened due to the degree of valleyward movement. The detached blocks of rock represented by the ridges have acquired an apparent dip downslope. This effect is seen further in profile D, where the trough is wider due to further movement, and the apparent dip of the beds in the ridge is greater. The forms are more smoothed and rounded, due presumably to an increased length of time during which surface processes have been operating. In the field, this ridge and trough have almost continuous soil and vegetation cover, with few exposures of bare rock, but the slope immediately below the ridge remains steeper than that further downslope. On profile E, further movement has displaced the ridge down the slope, and surface processes have distributed most of the steepening caused by its presence along much of the length of the slope, both above and below it, and are filling in the trough. In profile F the reversed slope on the upper side of the ridge has been eliminated almost entirely, and surface processes are the dominant movements on the slope as a whole.

In this context of mass wasting, it is clear that surface instability, such as that due to the exceeding of its plastic limit by the field moisture content of a soil, is chiefly of importance in the later stages. Such instability must play a significant role in the reduction of slope ridges formed by large-scale block glides, and assist in the re-establishment of the rectilinear slope form.

CONCLUSIONS

The slope soils of the Hambleton Hills have lower liquid limits and plastic limits, and smaller plasticity indices, than are generally obtained from clay materials during engineering site investigations, but compare with other surface soils (Gill & Reaves 1957; Soane et al 1972). The higher silt and sand content in surface loams is probably the significant factor. As commonly observed in soil mechanics investigations, the plasticity index of the slope soils in the Hambleton Hills is greater with higher proportions of clay. The liquid limit increases with the plasticity index, and the plastic limit increases with the liquid limit. Thus, the greater the clay content, the higher the field moisture content must rise before the soil becomes plastic and potentially unstable. Soils which have relatively low maximum field moisture contents also have relatively low liquid and plastic limits. Conversely, soils with high maximum field moisture contents have high liquid and plastic limits. Thus, whilst the attainment of potential instability depends on a high field moisture content being attained, the critical level varies from soil to soil.

It seems that occasional downpours, under the present climatic regime and even during an average year of unexceptional precipitation events, are sufficient to produce potential instability. The magnitude of the increment of precipitation, additional to that which produces the normal seasonal maximum needed to raise the field moisture content of a soil to the plastic limit, does not depend on the absolute level of the seasonal maximum. It depends on the magnitude of the difference between that maximum and the plastic limit. That the liquid and plastic limits should be significantly correlated with parameters of field moisture content and with particle size characteristics, confirms their importance as indices of soil behaviour. However, it can be seen that the relationship between actual field moisture conditions and standard soil mechanics tests for plasticity suggest that these tests over-estimate the susceptibility of natural, vegetation-covered ground surface to soil flow.

Nonetheless, surface movements by wash, creep and flow are important in the elimination of ridge-and-trough features caused by large-scale mass movement. The corrasive effects of the regolith in motion, suggested for slopes on the same rock outcrop 36 km to the east of Hambleton Hills (Palmer 1956) possibly play a significant part in this. While the impressive block glides in the area are large-scale events of low frequency, and soil creep and surface wash are small-scale events of high frequency, it would seem that soil flow falls somewhere in between, in terms of both scale and frequency.

ACKNOWLEDGEMENTS

Gareth Shaw and Helen Cooper assisted with field survey of slope profiles, and Martin Lee assisted with the particle-size determinations. The work was initiated whilst the author was supported by a Natural Environment Research Council Studentship and supervised by Dr. A. F. Pitty.

REFERENCES

Anderson, G.D., 1958. *A preliminary investigation of the soils of the north-east Yorkshire moors*. Unpublished PhD thesis, School of Agriculture, King's College, University of Durham

Atterberg, A., 1911. Die Plastizität der Tone. *Internationale Mitteilungen für Bodenkunde*, 1, 10-43

Baver, L.D., Gardner, W.H. & Gardner, W.R., 1972. *Soil physics*. (Wiley, New York), 4th edition

Bendelow, V.C. & Carroll, D.M., 1976. *Soils in North Yorkshire III: parts of Sheets SE 79, 88, 89, 98, 99 (Pickering Moor and Troutsdale)*. Soil Survey of England & Wales, Soil Survey Record 35

Bendelow, V.C. & Carroll, D.M., in press. *Soils in North Yorkshire IV: Sheet SE 58 and parts of Sheets SE 49 and 59 (Rievaulx and Upper Ryedale)*. Soil Survey of England & Wales, Soil Survey Record

Briggs, D.J. & Courtney, F.M., 1972. Ridge-and-trough topography in the north Cotswolds. *Proceedings of the Cotteswold Naturalists' Field Club*, 36, 94-103

British Standards Institution, 1967. *Methods of testing soils for civil engineering purposes*. (British Standard 1377: 1967, London)

Catt, J.A., Weir, A.H. & Madgett, P.A., 1974. The loess of eastern Yorkshire and Lincolnshire. *Proceedings of the Yorkshire Geological Society*, 40, 23-39

Cooper, R.G., 1977. Quarrying in the Hambleton Hills, North Yorkshire: the problems of identifying disused workings. *Industrial Archaeology Review,* 164-170

Cooper, R.G., 1978. The discovery and exploration of the North Yorkshire windypits. *Ryedale Historian,* 9, 10-12

Cooper, R.G., Ryder, P.F. & Solman, K.R., 1976. The North Yorkshire windypits: a review. *Transactions of the British Cave Research Association,* 3, 77-94

Fitton, E.P. & Mitchell, D., 1950. The Ryedale windypits. *Cave Science,* 2, 162-184

Folk, R.L. & Ward, W.C., 1957. Brazos River Bar: a study in the significance of grain size parameters. *Journal of Sedimentary Petrology,* 27, 3-26

Fox-Strangways, C., 1892. *The Jurassic rocks of Britain. Volume 1. Yorkshire.* Memoirs of the Geological Survey of the United Kingdom, Her Majesty's Stationery Office, London

Gill, W.R. & Reaves, C.A., 1957. Relationships of Atterberg limits and cation-exchange capacity to some physical properties of soils. *Proceedings of the Soil Science Society of America,* 21, 491-494

Hodgson, J.M. ed., 1974. *Soil survey field handbook. Describing and sampling soil profiles.* Soil Survey of England & Wales, Soil Survey Technical Monograph 5

Imeson, A.C., 1971. Heather burning and soil erosion on the north Yorkshire moors. *Journal of Applied Ecology,* 8, 537-542

Leopold, L.B., Wolman, M.G. & Miller, J.P., 1964. *Fluvial processes in geomorphology.* (Freeman, London)

Meteorological Office, 1963. *Averages of temperature for Great Britain and Northern Ireland 1931-60 in degrees Celsius.* Meteorological Office 735, Her Majesty's Stationery Office, London

Palmer, J., 1956. Tor formation at the Bridestones in north-east Yorkshire, and its significance in relation to problems of valley-side development and regional glaciation. *Transactions of the Institute of British Geographers,* 22, 55-72

Pitty, A.F., 1968. A simple device for the field measurement of hillslopes. *Journal of Geology,* 76, 717-720

Pitty, A.F., 1979. *Geography and soil properties.* (Methuen, London)

Savigear, R.A.G., 1952. Some observations on slope development in South Wales. *Transactions of the Institute of British Geographers,* 18, 31-51

Seed, H.B., Woodward, R.J. & Lundgren, R., 1964. Fundamental aspects of the Atterberg limits. *Journal of the Soil Mechanics & Foundation Division, Proceedings of the American Society of Civil Engineers,* 90(SM6), 75-105

Slope soils near Helmsley, N. Yorkshire

Skempton, A.W., 1953a. The colloidal 'activity' of clays. *Proceedings of the 3rd International Conference on Soil Mechanics & Foundation Engineering,* 1, 57-61

Skempton, A.W., 1953b. Soil mechanics in relation to geology. *Proceedings of the Yorkshire Geological Society,* 29, 33-62

Soane, B.D., Campbell, D.J. & Herkes, S.M., 1972. The characterization of some Scottish arable topsoils by agricultural and engineeering methods. *Journal of Soil Science,* 23, 93-104

Varnes, D.J., 1958. Landslide types and processes. In: *Landslides and engineering practice,* ed. E. B. Eckel, United States Highway Research Board, Special Report 29

CHAPTER 8 THE USE OF DIFFERING SCALES TO IDENTIFY FACTORS CONTROLLING DENUDATION RATES

R. R. Arnett[*]

INTRODUCTION

With the increasing importance of contemporary process studies in the context of drainage basin analysis, the transport of weathered debris by runoff has become a major focus for the collection and interpretation of numerical information. The scales of such measurements have varied spatially from 'at a point' studies of net erosion rates (Imeson 1974) to sediment budgets derived from the world's largest catchment areas (Holeman 1968).

Sediment-yield studies incorporate four major aims. Firstly, by determining the total amounts of suspended, dissolved and tractive material discharged from landmasses of any size, overall rates of denudation may be calculated (Strakhov 1964). From such rates conclusions may be inferred about long-term landsurface modification. Where definable time-lags exist between causal process and final form, such as in gully erosion, mass-movement features or channel and floodplain patterns, a study of sediment movement through the system will also facilitate explanation. Thirdly, a study of the hydro-meteorological and biopedological factors involved in sediment transport clarifies our understanding concerning inter-related cycles in the natural environment which result in organic and mineral debris being discharged from a drainage basin (Bormann & Likens 1969; Arnett 1978). Finally, sediment yield studies have a practical importance in the design of reservoirs, flood mitigation and soil erosion prevention schemes, and urban storm drainage systems (Leopold 1968).

THE RANGE OF SCALES INVOLVED

In discussing sediment yields on a world-wide basis, climatic variation is considered to be dominant (Fournier 1960), with a variety of other factors modifying the results. The climatic components are temperature and effective precipitation, the former controlling rates of chemical breakdown in the lithosphere and the latter providing the means for downslope transportation. High combinations of both factors should produce maximum rates of denudation. Strakhov's (1964) conclusions, based on data from 60 catchments, indicate that suspended sediment losses are zonal with the $10°C$ mean annual isotherm dividing rates of less than 20 tonnes/km^2 in higher latitudes up to 2000 tonnes/km^2 in

---
[*] Department of Geography, University of Hull

south-east Asia. Due to the size, world-wide distribution, and environmental complexity of these sub-continental basins, however, precipitation and temperature characteristics are the only definable parameters which account for some of the observed variation in sediment totals between them. Using more data from within the European sub-continent, where climatic variation is less extreme, Fournier (1960) claims that even within one climatic zone, sediment losses may vary from 20 to 6900 tonnes largely as a result of human interference, a conclusion strongly supported by Douglas (1967).

Reducing the size or latitudinal range of the catchment areas involved, lessens the control of macro-climate as precipitation and temperature variations are reduced. Maner (1958) in the Red Hills district of the United States, Diaconu (1969) in Roumania and Bauer & Tille (1967) in Thuringia all relate suspended sediment variations to pedological, lithological and land-utilisation factors, while a similar multivariate approach has been demonstrated by Anderson (1954), Wallis & Anderson (1965), Jansen & Painter (1974) and Glymph (1975). At the micro-scale of subcatchments or 'at a point' studies, the source areas for sediment become extremely localised (Bridges & Harding 1971), so that the range of published rates increases from zero, where dense vegetation prohibits erosion or even induces accumulation (Imeson 1971), to 40 000 tonnes/km^2 for actively eroding gully systems in the Howgill Fells, north-west England (Harvey 1974).

THE STUDY AREA AND SCALES OF STUDY

The main aim of this paper is to identify the differing importance of environmental variables in controlling suspended and dissolved sediment yields within the North Yorkshire Moors National Park, north-east England. A major aspect of the geographical element in the present approach, apart from studying a large number of cases in the study area, is to focus attention on the critical variables by altering the scale of the area investigated. Thus three scales of study are defined, the drainage basin, the subcatchment and the controlled plot experiment. In the first case, variation in annual sediment loads between 16 river basins is described and explained, while in the second, storm concentrations for particulate and solutional components in 93 subcatchments are similarly treated. At the most detailed level, eroded material from three controlled plots is considered, varying only in the amount and type of vegetation cover.

The North Yorkshire Moors are a range of Jurassic Hills bordering the North Sea between Middlesbrough and Scarborough, and bounded by the vales of York and Pickering to the west and south. Lithologically, the strata are horizontally bedded ranging from Lower Lias to Kimmeridge Clay, although considerable lateral variation in facies occurs in the estuarine deposits (Rayner & Hemingway 1974). The 16 drainage basins indicated in Figure 8.1 cover an area of 900 km^2 and range in size from 10 to 300 km^2. Whereas the solid lithology is dominant in the central, upland parts of the study area, the marginal basins to the east, north and

Factors controlling denudation rates

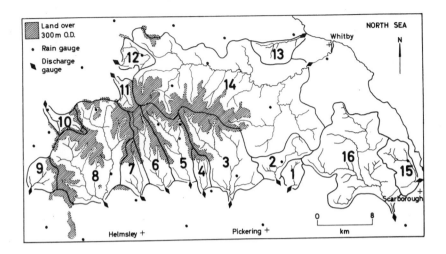

Figure 8.1   Map of the North Yorkshire Moors, showing the boundaries and position of gauging stations of the catchments studied. The numbering of the catchments corresponds to the numbers and names listed in Table 8.1.

west are wholly or partially plugged with boulder clay dating from the most recent Quaternary ice advances.

1.   Scale 1.   The drainage basin.

The determination of total sediment lost from the study area was based on the construction of rating curves relating the concentration of dissolved and suspended material to stream discharge at the time of sampling. By abstracting manual and automatic water samples from all streams over a period of 18 months from November 1974 to April 1976, a wide range of flow conditions was encountered, thereby enabling all sections of rating curves to be calculated. Having established sediment concentrations for the range of discharge conditions, annual runoff data were utilised to compute sediment loads. Eleven basins are monitored by regional water authorities using continuous stage recorders, while on the five remaining streams channel cross-sections were surveyed and stage recorders installed. From all graphical records for 1975, hourly stages were abstracted using a line digitiser, and by applying the relevant rating equations, hourly discharge and sediment loads were computed.

A method was also required whereby the physical components of each basin could be described and compared, and information concerning relevant environmental variables was accumulated on a grid basis to facilitate storage and analysis. National Grid Squares were divided into 9 units, each 11.1 hectares in area, giving 8 100 cells covering the 16 basins, for which topographical, lithological, drainage density and landuse indices were defined. From 1:25 000 maps, maximum and minimum altitude per square provided

Factors controlling denudation rates

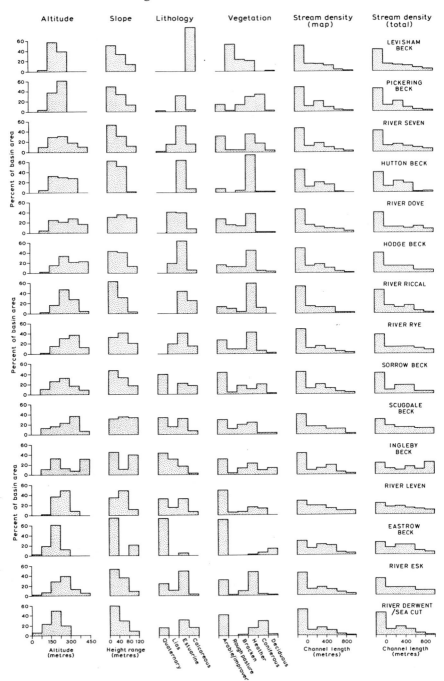

Figure 8.2 Drainage basin characteristics. Each horizontal set of histograms summarises the environmental characteristics of one particular basin. Intra-basin variation for one variable is revealed in each vertical column.

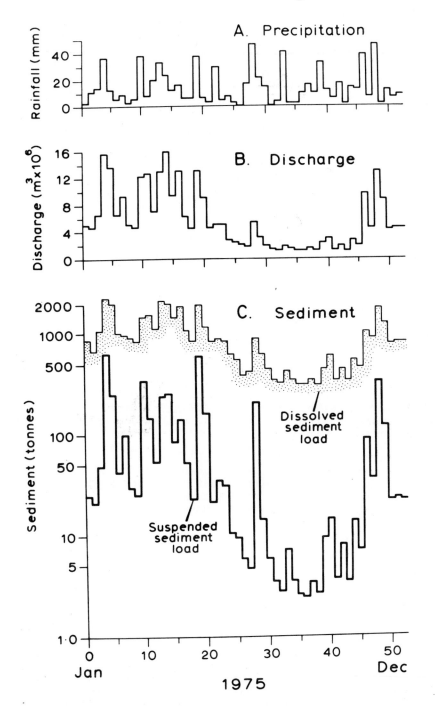

Figure 8.3   Weekly sediment losses from the study area as a whole.

Factors controlling denudation rates

Table 8.1 Regression equations for sediment rating curves. In all cases 'X' = discharge (m^3/second)(Log$_{10}$). Figures in brackets represent correlation coefficients.

		Suspended solids concentration (log$_{10}$ mg/litre)	Dissolved solids concentration (log$_{10}$ mg/litre)
1.	Levisham B	0.86X + 1.17 (0.73)	2.31 - 0.15X (-0.81)
2.	Pickering B.	0.71X + 1.10 (0.93)	2.10 - 0.14X (-0.85)
3.	R. Severn	0.45X + 0.65 (0.83)	2.12 - 0.20X (-0.87)
4.	Hutton B.	0.61X + 1.02 (0.86)	2.00 - 0.16X (-0.80)
5.	R. Dove	0.98X + 0.95 (0.85)	2.00 - 0.16X (-0.80)
6.	Hodge B.	0.69X + 0.76 (0.85)	1.09 - 0.18X (-0.80)
7.	R. Riccal	0.67X + 0.76 (0.84)	1.92 - 0.28X (-0.92)
8.	R. Rye	1.05X + 0.39 (0.86)	2.20 - 0.16X (-0.68)
9.	Sorrow B.	0.46X + 1.20 (0.72)	2.43 - 0.04X (-0.70)
10.	Scugdale B.	0.90X + 1.51 (0.86)	2.32 - 0.11X (-0.75)
11.	Ingleby B.	0.82X + 2.04 (0.87)	2.13 - 0.12X (-0.78)
12.	R. Leven	1.51X + 1.87 (0.86)	2.21 - 0.12X (-0.87)
13.	East Row B.	1.30X + 2.18 (0.91)	2.44 - 0.12X (-0.72)
14.	R. Esk	0.93X + 0.46 (0.87)	2.26 - 0.14X (-0.83)
15.	Sea Cut	0.36X + 0.77 (0.84)	2.39 - 0.25X (-0.94)
16.	R. Derwent	1.03X + 1.04 (0.74)	2.29 - 0.29X (-0.74)

indices for mean altitude and average slope, while lithology was coded from one-inch and six-inch geological maps. Dominant vegetation types were defined from 1:10 000 air photographs and the total length, number and magnitude of stream channels in each grid square were digitised from 1:25 000 maps. Such maps, however, omit low order, ephemeral gully systems which act as major sediment sources during storm conditions, and consequently total networks visible on air photographs were transcribed to 1:10 000 maps and likewise digitised. A summary of all these environmental components are presented as histograms in Figure 8.2.

Sediment losses for the entire area are shown in Figure 8.3, together with relevant discharge and precipitation information. Over the twelve month period, 52 100 tonnes were evacuated from the 16 basins giving an average loss of 589 kg/hectare. This figure excludes material removed from the drainage basins as bedload, which may increase the total by 10-15 per cent (Hall 1967). The sediment rating curves, each calculated from a minimum of 70 analysed water samples, are given in Table 8.1 and illustrated in Figure 8.4 where, although regression lines appear generally similar, marked variations occur in both slope and intercept terms, reflecting the different environmental conditions occuring in each basin. The accuracy of the equations relating observed sediment concentrations to discharge varies between particulate and solutional components, and between basins for each component. The amount of material in suspension increases with discharge, more particles being transported at higher stream velocities. Correlation coefficients range from r = 0.72 to 0.93, indicating that other factors are involved, notably antecedent climatic conditions and seasonal changes in the vegetation cover. With regard to dissolved sediment concentrations, the efficiency of the linear equations is slightly less, the coefficients ranging from r = -0.63 to r = -0.92. Here the basic dilution effect produced by flood discharges is complicated not merely by antecedent factors referred to previously, but also by the 'flushing' effects of solutes from the soil at the close of long dry periods, and the different responses of component ions to particular environmental conditions. Available evidence would indicate that for detailed solute budgets, rating curves should be replaced by rating 'loops'. However, for the purpose of this account, the stated equations are considered satisfactory for calculating annual sediment budgets. As the total study area covers only 900 km^2, climatic variation is unlikely to be a dominant factor. From Fig 8.4, basin 12 (Ingleby Beck) appears distinct in having the steepest regression line, reflecting a favourable combination of conditions for high sediment concentrations. The catchment is incised into the north-facing escarpment of the upland area, giving steep, unstable upper slopes covered with a dense network of gully systems, and a valley floored with unconsolidated glacial sediments (Fig 8.2). In contrast, basins 3 to 7, sub-parallel adjacent valleys draining south into the vale of Pickering, reveal low rates of sediment increase with discharge. Unaffected by direct ice or meltwater activity during the last major ice advance, and incised mainly into the relatively porous estuarine sandstone, sediment availability is greatly reduced. These basins also have a much smaller percentage of their area devoted to arable cultivation, largely limited to the Lias clay inliers of Farndale, Rosedale and Bransdale (Fig 8.2). Regarding the suspended solids equations generally, the slope values range from 0.3 to 1.5, considerably lower than other published figures which often exceed 2.0, implying a rapid acceleration in concentration as compared to discharge. One possible reason for the low figures derived from this study may be the unusual dryness of the year in question, 1975. The absence of severe storm events, which usually contribute the bulk of particulate sediment (Piest 1965) would inevitably bias the regression estimate towards low flow conditions.

Dissolved solids equations reveal less variation in both regression coefficients, indicating that solution concentrations react similarly to changes in discharge over the whole area. In this case, the differing environmental backgrounds to each basin (Fig 8.2) are far less effective in differentiating dissolved losses than particulate ones. Although the parallel nature of the regression lines is revealed by similar slope coefficients in Table 8.1 some relative displacement is evident reflecting varying availability of solutional material. Catchments with higher intercept coefficients fall into two categories, those such as Levisham Beck and the River Rye which drain calcareous

Factors controlling denudation rates

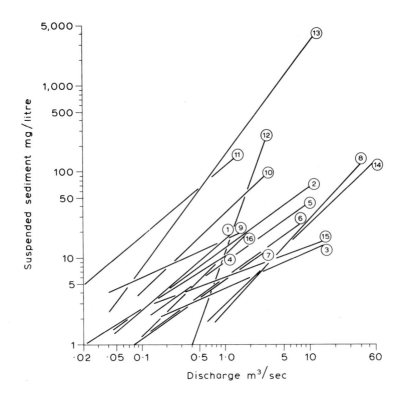

Figure 8.4A   Suspended sediment concentrations as related to discharge. See Table 8.1 for the key to numbers.

upper Jurassic gritstones, and the marginal basins (numbers 9 to 14) partially filled with glacial material. Valleys of the River Derwent and Sea Cut fall into both categories. As with suspended sediment, the southerly draining catchments from Rosedale (River Severn) across to Riccaldale (River Riccal) are distinctive in possessing smaller intercept terms, reflecting lower solute availabilities from the leached Jurassic sandstones.

Annual budgets for individual basins are given in Table 9.2 and illustrated in Fig 8.5. Total losses per hectare vary from 300 kg in Ingleby Beck to 1149 kg in Levisham Beck, an average of 579. Solution is the dominant transportation medium, accounting for 91 per cent of the total load, ranging from 1132 kg/hectare in the wholly calcareous Levisham Beck to 258 kg/hectare in Ingleby Beck. Figures given in Table 8.1 for solutional and total losses, however, include inputs from airborne salts (Cryer 1976). Sodium losses, for example, ranging from 5 to 55 kg/hectare, are likely to include sodium chloride from the adjacent North Sea. Suspended losses are small in comparison, with all basins except two losing less than 50 kg/hectare and varying from 17 to 320 kg/hectare in Scugdale and East Row Becks respectively. These variations displayed in Fig 8.5 have considerable geographical implications. With

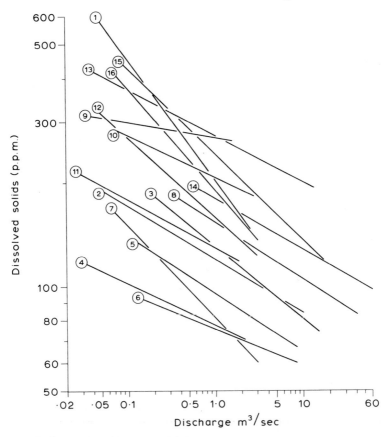

Figure 8.4B Dissolved solids concentrations as related to discharge. See Table 8.1 for the key to numbers

increasing use of models and equations, more importance is now being attached to average values and confidence limits than to residuals and deviant elements which are classified as 'noise'. A geographical rather than purely statistical approach demands just the opposite; the collection of data from a larger number of points in space from which generalisations may emerge, but only as a result of, not at the expense of, environmental complexity. In studying 16 basins, each with a unique environmental composition, sufficient permutations of independent variables exist from which conclusions may be drawn for the group as a whole for each individual basin. A mean suspended sediment loss of 47.5 kg/hectare for the entire study area has very limited value as none of the basins produced this amount in 1975 although displaying a range from 17 to 320. Such an average may be useful for comparison with other similarly derived values, but adds little to our knowledge about the factors promoting variable sediment yields within this upland zone.

Utilising the regional denudation rates in Table 8.2 as dependent variables, controlled by environmental indices

Factors controlling denudation rates

Table 8.2 Annual sediment budgets for 1975

Basin	Area (km²)	Suspended		Dissolved	
		total loss (tonnes)	rate (kg/hec)	total loss (tonnes)	rate (kg/hec)
1. Levisham B	11.1	18.9	17	1258.8	1133
2. Pickering B	24.2	109.0	45	1375.7	568
3. R. Seven	85.0	229.5	27	4224.5	497
4. Hutton B.	11.6	22.0	19	405.8	351
5. R. Dove	46.2	194.1	42	1955.1	423
6. Hodge B.	37.2	78.2	21	1287.8	346
7. R. Riccal	22.0	22.0	10	873.4	397
8. R. Rye	130.7	522.6	40	6938.1	531
9. Sorrow B.	18.8	46.9	25	1416.0	754
10. Scugdale B.	13.6	23.0	17	827.9	611
11. Ingleby B.	9.7	40.6	42	249.5	258
12. R. Leven	15.1	68.0	45	1072.8	710
13. East Row B.	19.7	629.4	320	1416.2	720
14. R. Esk	299.4	2066.2	69	14582.7	487
15/16 Sea Cut/ Derwent	155.8	342.7	22	10795.6	693

derived from Table 8.2 multiple regression analysis was undertaken. The Sea Cut and River Derwent were considered as one for this analysis, as runoff is artificially channelled from one to the other during flood. For suspended sediment, variations between the 15 basins is most satisfactorily explained by:

$$\text{Log}_{10} \text{ RSUS} = 0.53 \log_{10} \text{Agric} - 0.42 \log_{10} \text{Calc} + 1.102$$

(multiple correlation coefficient = 0.92)

where RSUS = rate of suspended sediment loss (kg/hectare)

Agric = per cent of basin under improved pasture or arable

Calc = per cent of basin occupied by calcareous rocks

At this level of analysis, the detailed indices of slope, altitude and stream density are disregarded in favour of agricultural practices and impervious lithologies. Most sediment loss, therefore, is not generated in upland basins with a large moorland or rough pasture component, but from catchments where the dominant landuse is arable or ley pasture. The negative effect of calcarous rocks implies that as the percentage of limestone in a basin increases, the rate of sediment loss declines, with more water reaching the channel as filtered underground seepage rather than turbid surface runoff.

For dissolved sediment, the amount of agricultural land is again the most important factor explaining interbasin

Factors controlling denudation rates

variation. As a strong association exists between the agricultural land and areas of Quaternary deposits, it would appear that being geologically recent and already in a partially weathered condition, the boulder clay provides a ready source of nutrients for water percolating slowly through the relatively impermeable material. The optimum regression solution is:

$$\text{Log}_{10}\text{RDIS} = 0.67 \text{ Log}_{10}\text{Agric} - 0.96 \text{Log}_{10}\text{Totdens} - 3.31$$

(multiple correlation = 0.88)

where RDIS = rate of dissolved solid loss (kg hectare)

Agric = per cent of basin under improved pasture or arable

Totdens = total stream density (m/hectare)

The second variable emerging from this analysis is average total stream density. As channel length declines, the rate of dissolved sediment loss increases, presumably by lengthening transmission time for water moving through the soil as compared with open channel networks.

Two general conclusions emerge from this survey of regional denudation rates, demonstrating that losses from one catchment are not an accurate guide to losses from adjacent basins, and certainly would have little meaning if entered into a table listing several isolated catchments each from different regions. For example, calculating losses for the whole study area on the basis of individual sediment yields produces a figure which ranges from 27 000 tonnes using results from basin 11, to 103 500 tonnes based on basin 1. Discrepancies for suspended solids alone are even greater, minimum and maximum estimates being 1530 and 28 800 tonnes respectively, depending on the catchment used to provide the experimental data. The second conclusion suggests that at this level of interbasin analysis, within one climatic regime, sediment losses are more closely related to lithology and landuse factors rather than to measures of relief, slope or drainage density.

2. Scale 2. The subcatchment.

Sediment yields from large catchments only indicate net losses, revealing little about source areas or resulting changes in specific landforms. Eroded material may originate from permanent channels, ephemeral gully systems or adjacent slopes, while denudation in some areas may be counterbalanced or outweighed by deposition elsewhere. To investigate how the variation noted in Fig 8.5 would change at a larger scale, 93 subcatchments, ranging in size from 0.9 to 50 km^2, were defined inside the 16 major basins. The compilation of rating curves for so many streams was inconceivable, and consequently storm concentrations were employed for suspended and dissolved sediment. As spatial variability in rainfall patterns eliminated the necessity for 93 simultaneous discharge and sample measurements, Fig 8.6 displays the averages of two concentrations from the falling limb of different storm hydrographs. However, by utilising large numbers of personnel and equipment, all subcatchments within each of the major basins were sampled

Factors controlling denudation rates

Table 8.3 Suspended sediment concentrations in storm runoff, mg/l

Basin	Suspended		Dissolved	
	Main basin	Subcatchment range	Main basin	Subcatchment range
1. Levisham B.	19	14 - 106	144	243 - 267
2. Pickering B.	36	86 - 128	108	71 - 120
3. R. Seven	26	27 - 333	82	74 - 109
4. Hutton B.	29	34 - 59	73	61 - 65
5. R. Dove	43	152 - 401	77	64 - 110
6. Hodge B.	23	85 - 308	65	58 - 112
7. R. Riccal	24	74 - 152	64	76 - 102
8. R. Rye	78	103 - 412	97	75 - 214
9. Sorrow B.	26	148 - 247	286	194 - 310
10. Scugdale B.	68	275 - 306	163	94 - 210
11. Ingleby B.	114	532 - 859	139	122 - 249
12. R. Leven	145	124 - 231	126	86 - 238
13. East Row B.	1260	153 - 597	153	212 - 267
14. R. Esk	98	82 - 280	119	74 - 204
15/ Sea Cut/ 16 Derwent	24/16	13 - 300	110/158	65 - 257

during the same storm event. Table 8.3 reveals the range of concentrations within each large basin, together with results from concurrent samples for the main basin outlets. For example, the river Rye produced a mean particulate concentration of 78 mg/litre, although 11 subcatchments ranged from 103 to 412 mg/litre. Consequential upon the lack of control over antecedent meteorological conditions, any comparison of results over the entire area should be treated cautiously, but as Fig 8.6 and Table 8.2 indicate, the range of sediment concentrations is much greater at the subcatchment level than between major basins. Sediment source areas therefore appear to be extremely variable within large watersheds, with erosion in some areas being compensated by deposition in others. In terms of the River Rye, the headwater tributaries yield large quantities of suspended sediment, but a large portion of this must be deposited in the lower valley, to explain low net concentrations at the main basin outlet.

As with the major basins, each subcatchment may be defined by identical environmental indices to those shown in Fig 8.2 but, in this instance, multivariate analysis is restricted by the inconsistent nature of the sediment concentration data. However, the pattern of particulate values in Fig 8.6 closely reflects the distribution of two environmental variables, mean relative relief and mean total channel density. Maximum suspended sediment concentrations

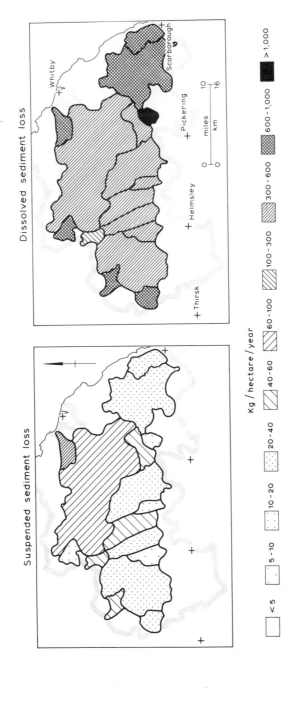

Figure 8.5   Average total sediment losses from the North Yorkshire Moors. (Boundary of National Park indicated by dense dots)

Factors controlling denudation rates

are spatially associated with subcatchments possessing steep relief (defined as the mean altitude range of all grid squares in the basin), and a high channel density (total channel length divided by basin area). The steep slopes indicate skeletal soils with low infiltration capacities while the dense channel networks, incorporating tributary gully systems, ensure that sediment is speedily evacuated from such catchments in flood conditions. Multiplying mean relief (metres per grid) by total channel density (metres per hectare) produces a density-relief index (Fig 8.7) which coincides closely with the particulate concentration map (Fig 8.6). Within the general similarity, however, distinct anomalies occur between the two patterns. Suspended sediment concentrations in East Row Beck subcatchments do not correspond with either steep relief or high drainage densities, but confirm the significance of lithology. This entire basin is filled with boulder clay, while the storm concentrations for the catchment as a whole exceed sub-catchment yields by at least a factor of two (Table 8.3) suggesting that the channels themselves provide most of the load rather than adjacent slopes. The density-relief index on the other hand emphasises certain subcatchments particularly in Levisham Beck, the lower Derwent and lower Rye valleys, which yield relatively low particulate concentrations. A possible explanation for such discrepancies may be the fossil nature of many channel systems particularly along the southern margins of the study area. In these permeable, calcareous sandstones of the Upper Jurassic, high channel densities appear to reflect periglacial conditions towards the end of the Quaternary period, greatly reducing the permeability of the regolith and increasing surface runoff.

Whereas variation in suspended sediment yield between the 16 major basins was largely attributed to the presence of agriculturally utilised boulder clay, study of smaller scales suggests that within basins the supply of eroded material is partially controlled by a relief and drainage density factor. With dissolved material on the contrary the subcatchment analysis merely confirms the dominance of lithology and landuse as causal factors. As the sub-catchments are much smaller in size, variation in the environmental variables is more controlled, and the difference between calcareous and non-calcareous influences easier to detect. Storm concentrations exceeding 200 mg/litre are exclusively limited to the calcareous grit-stones of the Tabular Hills across the southern margin of the study area or to the unleached, agriculturally dominated boulder clay in the basins adjacent to the vale of York, the Tees plain or the North Sea. At the other extreme, the subcatchments in the central upland area dominated by heather and estuarine sandstones all record solutional concentrations less than 100 mg/litre.

3. Scale 3. The experimental plot

Reducing the scope of investigation to localising actual source areas for sediment away from the channel network itself allows a measure of control to be exercised over the multiplicity of environmental variables involved.

Factors controlling denudation rates

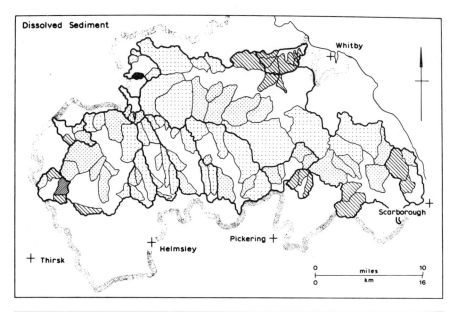

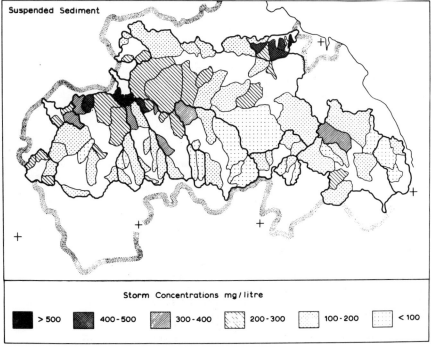

Figure 8.6 Sediment concentrations for the falling limb of storm hydrographs (average of two storms)

Factors controlling denudation rates

On the flanks of Westerdale Moor (grid reference 668012) close to the junction of the Esk, Dove and Seven basins, three experimental plots were defined, draining into two-metre wide, covered troughs, directing water and sediment into polythene storage tanks. The plots, 20 metres in length and extending up to the main water divide, were similar in terms of slope (15 degrees), lithology (estuarine sandstone) and aspect (west facing), varying only in the developmental status of the heather vegetation *(Calluna vulgaris)*. The *Calluna* in plot 1 was dense and 'mature' (Watt 1955) in stage, with a mean height of 43 cm above an organic layer 12 - 18 cm thick. In plot 2 the 'pioneer' community was shorter and less dense, growing in a thin layer of peat. The heather in plot 3 had been burnt 18 months previously and new *Calluna* shoots were barely appearing through a thin, discontinuous organic layer.

Water and sediment from the plots was measured over a period of 12 months between July 1972 and June 1973. Runoff volumes were recorded to the nearest 10 ml, sediment losses were ascertained by filtration and organic fractions determined by ignition. The results are given in Table 8.4, based on four-week totals. As anticipated, the highest runoff and sediment yields derive from freshly burnt *Callunetum*, amounts decreasing substantially through the 'pioneer' and 'mature' communities. In plot 1 the last two components are almost negligible even with rainfall totals exceeding 90 mm, and completely absent below 50 mm. A less dense vegetation cover and a thinner peat horizon in Plot 2 produces a general linear increase of runoff with rainfall, whereas in the plot devoid of *Calluna* the increase in runoff accelerates as precipitation amounts rise, until discharges from plot 3 exceed those from plot 1 by 16 times during maximum precipitation periods.

With regard to sediment removed by this runoff, similar relationships exist between the three plots. Little erosion is evident from the densely vegetated area with sediment loss increasing as the *Calluna* cover disappears. These observations relate to periods of the year when recorded precipitation was in the form of snow, lacking the direct erosive capability of raindrop impact. Most sediment from all plots is organic in origin (Table 8.4) although, on the recently burnt area, erosion was sufficient to penetrate the peat horizon and remove mineral soil. Under mature *Callunetum* runoff and sediment losses are small with a net accumulation of litter from growing plants (Imeson 1971). As the vegetation cover is destroyed, erosion is induced by successively lower rainfall amounts, with the result that denudational processes on recently burnt areas become more continuous throughout the year instead of occurring sporadically during high or heavy precipitation periods.

All sediment collected in the troughs was probably transported by overland flow, as there was no indication of gullying or micro-channel formation at the end of the observation period. Such a process may well account for the dominance of organic matter in the total sediment yield, having a lower specific gravity than mineral particles and

Factors controlling denudation rates

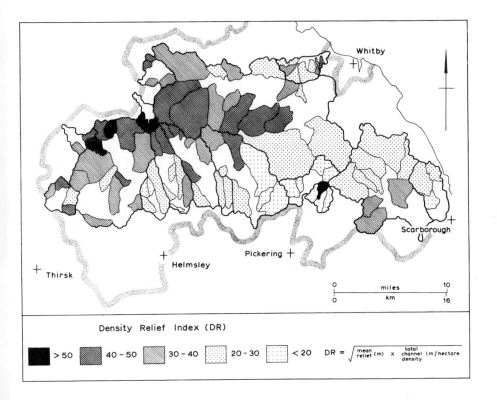

Figure 8.7 Map of density-relief indices for the North Yorkshire Moors

hence being more susceptible to transportation by overland flow. Where the peat layer has been removed to expose mineral soil, freeze/thaw action in the regolith rapidly loosens mineral grains in preparation for later erosion. Even following controlled burning in which only the upper woody portions of the plant are ignited, exposure of the organic layer to wind, rain and more extreme temperature variations (Barclay-Estrup 1971) inevitably promotes accelerated erosion. Where the burning is too severe, partially destroying the organic fraction, erosion is further increased and extended, the former because the removal of peat lowers the overland flow threshold, and extended due to slower regeneration of *Calluna* from seed (Whittaker & Gimingham 1962).

Even within subcatchments, therefore, denudation rates are highly localised depending upon specific combinations of topographic, landuse or hydrometeorological conditions (Arnett 1976). This plot experiment has demonstrated one set of conditions, namely the effect of natural or man-induced heather burning in promoting accelerated sediment loss but many other situations could produce similar effects.

Table 8.4 Annual mean discharge and sediment yield from three experimental plots on Westerdale Moor

Vegetation	run-off (litres)	total sediment (grams)	organic sediment (grams)
'mature' Calluna	17.3	7.9	7.0
'pioneer' Calluna	98.9	63.7	55.4
burnt Calluna	279.2	164.1	122.7

TEMPORAL VARIATIONS

The main aim of this study has been to examine the geography of sediment yields at contrasting scales. However, it must be noted that annual average figures for denudation rates mask significant variations throughout the year. The weekly distribution of dissolved and suspended sediment loads is displayed in Figure 8.3 reflecting the seasonal distribution of runoff, producing a variation in total losses from 303 to 2892 tonnes per week. This seasonality in discharge is accentuated and reduced by suspended and dissolved sediment respectively (Figure 8.3). In the former case, the rating curves are positive (Fig 8.4) giving maximum concentrations at peak discharge, while for dissolved sediment a dilution effect produces high concentrations during low flow conditions. Despite this dampening of the seasonal effect, however, solutional losses remain highest in winter, attaining a maximum of 2280 tonnes per week in late January, falling to 300 tonnes in early September. This dissolved budget of 47 700 tonnes represents 90 per cent of the total annual load, a far greater proportion than the 20-70 per cent quoted in the literature. The main reason for this discrepancy appears largely due to atypical precipitation inputs during the study year 1975. Annual total throughout the National Park amounted to only 75 per cent of the long term average, with major deficiencies occurring during the winter months. In February, for example, most gauges in the area recorded less than 40 per cent of their average amounts, with the result that runoff and sediment generated in succeeding wetter months were greatly reduced by high antecedent soil moisture deficits.

GENERAL CONCLUSIONS

Continental and regional denudation rates stated in the literature are usually based on a limited number of large catchment areas, and related to 'average' climatic conditions established over long time periods. Consequently, such rates are highly generalised both spatially and temporarily, offering little assistance to the geomorphologist attempting to relate sediment budgets in defined, spatially variable catchments to specific climatic conditions This paper has shown that in a small area of only 900 km^2 sediment yield for particulate and dissolved matter is highly variable, this variation increasing as the scale of investigation becomes more detailed, with the net yield from a basin of any size incorporating many nested phases of spatially distinct erosion and deposition. The greater the variability between a large number of sampling points, the easier it becomes to identify the controlling factors involved. Further, the study also demonstrates that attention can be focussed on particular factors by altering the scale of the study. Thus the influence of lithology and landuse are established at the inter-basin level, through slope and drainage density at the intra-basin scale down to an infinite combination of topographical, pedological and landuse components at the micro-scale.

ACKNOWLEDGEMENTS

The author gratefully acknowledges support from the Natural Environment Research Council who funded the main inter-basin project, and the Nawton Towers Estates for permission to site the experimental plots on Westerdale Moor. Discharge and rainfall information were kindly provided by the Yorkshire and Northumbrian Water Authorities, and the maps and diagrams drawn by the cartography staff, Hull University, Department of Geography. The manuscript was typed by the secretarial staff of the Department of Geography in the University of Queensland during the author's tenure of a Visiting Lectureship.

REFERENCES

Anderson, H.W., 1957. Relating sediment yield to watershed variables. *Transactions of the American Geophysical Union,* 38, 912-914

Arnett, R.R., 1976. Pedological factors influencing the permeability of hillside soils in Caydale, Yorkshire. *Earth Surface Processes,* 1, 1-14

Arnett, R.R., 1978. Regional disparities in the denudation rate of organic sediments. *Zeitschrift für Geomorphologie,* Supplement Band 29, 169-179

Barclay-Estrup, P., 1971. The description and interpretation of cyclical processes in a heath community. *Journal of Ecology,* 59, 143-156

Bauer, L. & Tille, W., 1967. Regional differentiations of the suspended sediment transport in Thuringia and their relation to soil erosion. *Publication of the International Association of Scientific Hydrology,* 75, 367-377

Bormann, F.H. & Likens, G.E., 1969. The watershed ecosystem concept and studies of nutrient cycles. In: *The ecosystem concept in natural resource management,* ed. G. M. Van Dine

Bridges, E.M. & Harding, D.M., 1971. Micro-erosion processes and factors affecting slope development in the Lower Swansea valley. In: *Slopes: form and process,* ed. D. Brunsden, Institute of British Geographers, Special Publication 3, 65-80

Factors controlling denudation rates

Cryer, R., 1976. The significance and variation of atmospheric nutrient inputs in a small catchment system. *Journal of Hydrology*, 29, 121-137

Diaconu, C., 1969. Resultats de l'étude de l'ecoulement des alluvions en suspension des rivières de la Roumanie. *Bulletin of the International Association of Scientific Hydrology*, 14, 51-90

Douglas, I., 1967. Man, vegetation and sediment yield of rivers. *Nature*, 215, 925-928

Fournier, R., 1960. *Climat et érosion: la relation entre l'érosion du sol par l'eau des precipitations atmospherique.* (Presses Universitaires de France, Paris)

Glymph, L.M., 1975. Evolving emphasis in sediment yield predictions. In: *Present and predictive technology for predicting sediment yields and sources.* United States Department of Agriculture, Agricultural Research Service Publication, ARS-S-40

Hall, D.G., 1967. The pattern of sediment movement in the River Tyne. *Publication of the International Association of Scientific Hydrology*, 75, 117-142

Harvey, A.M., 1974. Gully erosion and sediment yield in the Howgill Fells, Westmorland. In: *Fluvial processes in instrumented watersheds,* ed. K. J. Gregory & D. Walling, Institute of British Geographers, Special Publication 6, 45-58

Holeman, J.N., 1968. The sediment yield of major rivers of the world. *Water Resources Research*, 4, 737-747

Imeson, A.C., 1971. Heather burning and soil erosion on the North Yorkshire Moors. *Journal of Applied Ecology*, 8, 537-542

Imeson, A.C., 1974. The origin of sediment in a moorland catchment with particular reference to the role of vegetation. In: *Fluvial processes in instrumented watersheds,* ed. K. J. Gregory & D. Walling, Institute of British Geographers, Special Publication 6, 59-72

Jansen, J.M. & Painter, R.B., 1974. Predicting sediment yield from climate and topography. *Journal of Hydrology*, 21, 371-380

Leopold, L.B., 1968. Hydrology for urban land planning - a guidebook on the hydrologic effects of urban landuse. *United States Geological Survey, Geological Survey Circular*, 554

Maner, S.B., 1958. Factors affecting sediment delivery rates in the Red Hills physiographic area. *Transactions of the American Geophysical Union*, 39, 669-675

Piest, R.F., 1965. The role of the large storm as a sediment contribution. *Federal Inter-Agency Sedimentation Conference Proceedings, United States Department of Agriculture Miscellaneous Publication*, 970, 311-328

Rayner, D.H. & Hemingway, J.E., 1974. *The geology and mineral resources of Yorkshire*. (Yorkshire Geological Survey Society, Leeds)

Strakhov, N.M., 1967. *Principles of lithogenesis*. Translated by J. P. Fitzsimmons (Consultants Bureau, New York), Volume 1

Watt, A.S., 1955. Bracken versus heather, a study in plant ecology. *Journal of Ecology*, 43, 490-506

Wallis, J.R. & Anderson, H.W., 1965. An application of multivariate analysis to sediment network design. *Publication of the International Association of Scientific Hydrology*, 67, 357-378

Whittaker, E. & Gimingham, C.H., 1962. The effect of fire on the regeneration of *Calluna vulgaris*. *Journal of Ecology*, 50, 815-822

CHAPTER 9 WATER LEVELS IN PEATLANDS AND SOME IMPLICATIONS FOR RUNOFF AND EROSIONAL PROCESSES

R. W. Tomlinson[*]

INTRODUCTION

One of the main contrasts between the semi-arid type areas for fluvial process studies and the British Isles is the presence of peatland. Although peatlands cover only some 6.6 per cent of the United Kingdom (Robertson & Jowsey 1968), the distribution of this small proportion of land surface far outweighs its areal significance. With the exception of some valley mires, the majority of the peatland occupies the major upland watersheds of Britain, thus influencing the pattern of stream flow and erosion in the headwaters of most of the major streams.

Peatlands may differ in vegetation, structure and morphology. In consequence classification has proved difficult, but for considerations specifically of peatland hydrology, a classification based on relation to the general watertable within peat may be most suitable (Moore & Bellamy 1974). For peatlands, the water-balance may be expressed simply as

Inflow = Outflow + Retention

and peat growth may begin in the water retained in any depression, the peat "acting as an inert body displacing its own volume of water". Such peat is referred to as 'Primary Peat' and the mire so formed as a 'Primary Mire System'. As the peat develops beyond the confines of the basin it retains water and thereby becomes 'Secondary Peat' and the developing mire a 'Secondary Mire System'. Continued growth, such that it develops beyond the limits of the general ground watertable is referred to as tertiary peat growth. A 'Tertiary Mire System' is thus one in which the peat holds a volume of water above the general groundwater level. Clearly, such a tertiary mire can only obtain water from rainfall. Blanket mire, whereby peat has extended out from the basins and depressions to cover upland slopes, even those in excess of 25 degrees in more western areas of Britain, comprises the largest proportion of peatland in Britain. In terms of the above classification, blanket mire is Tertiary peat and, as such, obtains its water solely from interception of rainfall. Some raised and valley mires will occur, and have occurred, within the overall blanket area, thus complicating the simple inflow side of the equation. Further, even in the Tertiary blanket peat, changes have taken place in the hydrology as

[*] Dept of Geography, The Queen's University of Belfast

Water levels in peatlands

a result of both short- and long-term changes in climate. For instance a series of dry summers produces short-term effects whilst, at the other extreme, changes in climate are known to have occurred in post-glacial time, and according to some researchers, have produced vegetational and perhaps hydrological responses. Peatlands are, geographically, the starting point to many fluvial processes in upland Britain. The main purpose of the present account, therefore, is to describe measurements of water-table behaviour in a peat bog and then to consider some directions in which steps towards a greater understanding of water movement in and from peat areas might be taken.

THE STUDY AREA

As peatlands are very susceptible to even small environmental changes, a prerequisite for a study of actual hydrological phenomena in peat is that the study area should be remote from areas of human usage. As tertiary mires obtain mineral inputs only from rainfall, remoteness from atmospheric pollution is also desirable. These conditions are well-satisfied by mires lying within the hills of Galloway (Fig 1.9), with the Silver Flowe having the added advantage of being a National Nature Reserve. Here the Brishie Bog is one of a series of some seven distinct bogs which occupy the valley of the Cooran Lane and which are known collectively as the 'Silver Flowe' (Ratcliffe & Walker 1958).

1. Geology and relief

The valley of the Cooran Lane is some 6 km long and 1 km wide and trends approximately north-south (Fig 9.2). It is bounded on the east by the Kells Range and on the west by a line of hills which reach the maximum height in Craignaw at 645 m OD. Both the valley and the bounding hills are located within the Loch Dee granite mass but it is the metamorphosed surrounding rocks which now form the highest peaks of Merrick and Corserine. Glacial activity has greatly modified a valley which lies close to a former centre of ice dispersal (Pringle 1948) so that, on the west, slopes have been truncated and, especially on the eastern side, lateral moraines occur.

2. Climate and weather

The prevailing winds are from the west and south, bringing moist, mild air to the area. Rain falls on about 260 days a year and annual totals are between 1750 and 2000 mm for the region as a whole. Brishie Bog is in a slight rain shadow of the western hills but still received 1550 mm of rain during the study year of 1969. Relative humidity recorded in that year was also high, an average of 77 per cent, with lowest levels in May-June at 65-69 per cent, and reflected the seasonal distribution of rainfall. Maximum temperatures recorded in July were only 13.2°C whilst the maximum in January was 3.2°C. Frosts are frequent and may occur as late as the first week in June and as early as the second week in August although they do not occur regularly until September. Snow on the other

Figure 9.1 Pool-hummock complex at the centre of the Brishie Bog, with slopes of the Western Hills in the background.

hand is rare, only occurring on about 20 days in the year at a maximum. Sunshine is also limited, and during the period June 1968 to June 1969, the average daily length of sunshine was only 3.27 hours.

3. Peat

Under these prevailing climatic conditions, peatland is an inevitable feature of this part of Galloway. The Brishie Bog itself lies mid-way in the sequence lining the valley floor. Craigeazle, at the southern extremity, has all the features of a raised bog with its convex cross-section achieved by the almost flat central part and the very steeply sloping margin, or 'rand'. The Cooran Lane and its tributaries loop around the rand, giving an area of 'lagg', an area of groundwater movement with this flow ensuring a better nutrient supply and hence a richer vegetation. Round Loch and Long Loch bogs, at the northern end of the valley, have more the appearance of blanket bog although still possessing pool and hummock complexes at their centre. Brishie Bog contains elements of both. It has a convex outline with a rand and lagg on the south, east and west, but it continues northward in a broad sweep leading to blanket peat in the upper bogs. Above the rand, which rises steeply from the Cooran Lane on the south and eastern margins and from the Brishie Burn on the west, the centre of the bog is almost flat. The general fall is from northeast to southwest but an important variation is a slight depression in the centre-east of the bog. Some large pools lie within this depression, but others lie outside and have

Water levels in peatlands

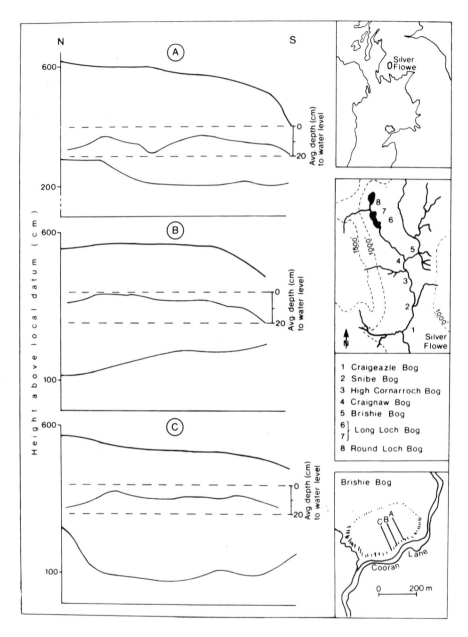

Figure 9.2 Location maps of the study area and cross-sections showing peat surface, average water levels and the form of the underlying mineral ground.

Figure 9.3 Continuous waterlevel recorder at a pool in the Brishie Bog (Site B_3)

long axes parallel to the contours. Nearly all pools lie in the eastern part of the bog, in contrast to the drier surfaces to the west which carry less *Sphagnum* but more *Rhacomitrium lanuginosum, Pleurozia purpurea* and *Cladonia uncialis* (nomenclature following Clapham, Tutin & Warburg (1962) for flowering plants, Richards & Wallace (1950) for mosses, and James (1965) for lichens). A stratigraphical cross-section indicated that this division has long been maintained and that the drier, western part of the bog lies above higher mineral ground.

MEASUREMENTS OF WATERTABLE OBSERVATIONS

The depth to water table in the wetter eastern part of the bog was examined in detail for a period of 18 months. Shallow wells were sunk along 6 transects which had an approximately north-south trend. At 10 m intervals a core of peat was removed to a depth of 50-60 cm by a sharpened piece of aluminium tubing slightly in excess of 5 cm in diameter. A piece of plastic fall-pipe tubing, 46 cm long, was placed in each well. Tests in both open- and Sphagnum-covered pools showed that drill-holes and the cavity at the base of the pipe allowed water levels inside and outside the pipe to equalise. Water levels were measured with an electrical probe (Goode 1970) which ensured that a distance of some 2 m could be maintained between the observer and the well, thus minimizing the effect of the observer's weight on watertable height.

Examination of the histograms for 12 months of measurement (Fig 9.4) shows that there is little variation in depth to watertable, which lies mainly between 0-10 cm

Water levels in peatlands

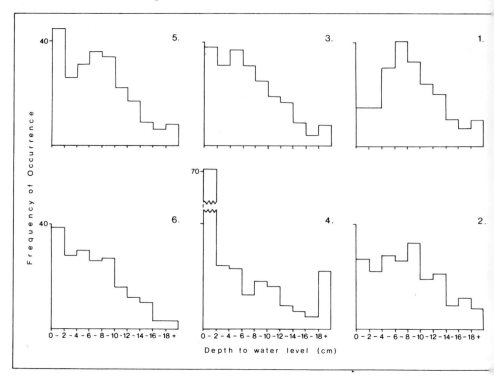

Figure 9.4   Histograms showing both consistency and some contrasts in the observed depths to water level in the Brishie Bog

depth. Individual transects vary, however, according to their position within the bog. In Transect 1, in the extreme east of the bog, depths of 6-8 cm are most frequently observed. Here the peat surface is rather like that of the western part of the bog, being relatively uniform with few hummocks and fewer pools. Towards the centre of the bog, in Transects 2-4, the depth to watertable decreases until, at Transect 4, there is a marked predominance of 0-2 cm depths, a clear expression of the abundance of pools in this area. A secondary mode in the 18 cm or more class is due to the rand conditions. Transects 5 and 6 are slightly to the west of the central depressed area of water collection but the influence of the pools is still evident although the modal class is not so distinct and the influence of the general mire expanse the relatively flat non-pool surface with *Sphagnum* spp., *Eriophorum angustifolium* and *Erica tetralix* is shown.

The degree to which seasonal variations depart from the average watertable level is a critical feature in the hydrology of peat bogs. In early summer, there is a marked drop in level in the present area. For example, in June 1969 on Transect 4 the height of water in the pools was between 5 and 7 cm below the average height whereas, in extremely wet months, several pools can be joined by the

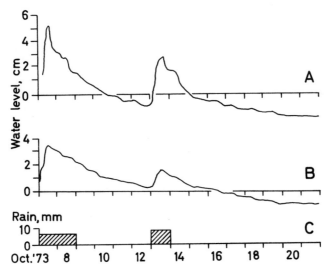

Figure 9.5   Examples of continuous water level records from the Brishie Bog after two rainfall events. Levels are given in relation to the 'critical level' described in the text.
A. Record from the mire expanse (site B_2)
B. Record from a surface pool (Site B_3, see Fig 9.3)
C. Precipitation recorded at Craigenbay (Detail from Boatman & Tomlinson 1973, courtesy of the British Ecological Society)

flooding of the central area of the bog. Such seasonal trends and other more detailed movements of the watertable, evident in the recorded charts (Fig 9.5), are an important indication of the hydrological characteristics of peatlands. When the watertable is relatively high, as in autumn (Fig 9.5), the watertable rises almost immediately when rain falls and drops rapidly when rainfall ceases. These two patterns indicate that flow through peat is slow. Otherwise watertable rise would not be so rapid. Even if rapid flow away is indicated after rain, the watertable only falls rapidly to a 'critical' level. Thereafter, water flow through peat almost ceases.

Falls in watertable during drier periods were not only slower but proceeded in regular pulses, producing a stepped effect in the record of both pools and of the mire expanse. As each drop in these 'steps' occurs during daylight hours, with the accompanying 'tread' at night, this effect is attributable to the key role of evaporation in peatland hydrology. That this role should be detectable in water-table fluctuations in the present area is not surprising, since potential evapotranspiration is 90 per cent and 33 per cent in June-July and in September, respectively (Boatman & Tomlinson 1973). Actual measurements of evaporation, using a Colorado evaporation pan, gave the

lower values of 75 per cent for a July rainfall of 93 mm
and only 6.6 per cent of a December precipitation of 197 mm.
That a diurnal effect is discernible in watertable fall
during summer emphasises the delicacies of water balance in
peatlands. The stepwise fall, however, is only noted in
summer below a certain level, but the depth of each step
appears to be independent of depth below this 'critical'
level, suggesting that the hydraulic conductivity of the
peat is negligible below that level. On Brishie Bog this
critical value appears to lie at 5-6 cm below the peat
surface. Chapman (1965) found a similar pattern at Coom
Rigg, with a critical level at about 8 cm. Runoff from the
bog increased sharply when the watertable was nearer the
peat surface. Similarly, Robertson et al (1968) found that
no movement of water occurred until the watertable was
established at a high level.

DISCUSSION

1. Relative impermeability of peat

The continuous water-level recordings suggest that only
a few centimeters beneath the surface of the Brishie Bog,
the hydraulic conductivity is so low that peat will impound
water once a pool is formed. Any changes in the water
level below the 'critical' depth can be attributed to
evaporation alone, with any lateral movement of water
probably being restricted to the upper 6 cm of the peat.
Thus, in southern Pennine moorlands, Tallis (1973) showed
that, once the watertable fell to 30 cm and more near to
a gully, flow in that gully ceased. Nearer the centre of
the bog, falls in the watertable in drier periods were less
marked, and, moreover, at a slower rate than falls near the
marginal gully. In addition to the hydraulic gradient
being low, it seems that the hydraulic conductivity of
peat is also low and that it also decreases with depth.
Although the measurement, and indeed the application of
Darcian hydraulic conductivity is suspect in peats,
values may be used for comparative purposes (Rycroft et al
1975). Tallis quotes values of 5.2×10^{-4} cm/sec for surface
peats, 37.2×10^{-4} cm/sec for Sphagnum but only 1.3×10^{-4}
cm/sec for peat from lower in the profile. These laboratory-
obtained values may be over-estimates, but confirm the scale
of decline in conductivity with depth. Hydraulic conductivity may also vary laterally. From these observations of
watertable height and limited conductivity, it appears that
the storage capacity of peats for water is low. In fact
Chapman (1965) gives a value of only 5 cm of rainfall.

2. Factors contributing to the relative impermeability of peat

Contrasted bog vegetations differ in the degree to
which they are conducive to water movement. Living
Sphagnum and other such unconsolidated material has higher
rates of flow (Daniels & Pearson 1974) than areas of
decomposed or herbaceous peats (Boelter 1965). This seems
to be the result of pore size, since the decomposed and
herbaceous peats have small pores which retain water.

Rycroft (1971) quotes Volarovitch and Churaev as recording that as much as 30 per cent of total water content in peat in the middle stages of humification is physically bound.

Decreased permeability of peat with depth is consistent with the development of peatlands in post-glacial times in Britain. The development of peat in the lower-lying areas was a response to increased wetness of the climate of the Atlantic Period. It was a primary peat, composed of herbaceous material and remains of carr woodlands, which became humified and compact. A subsequent phase was the 'lower peat' which began to form around 3000 BC or perhaps earlier. This is also humified, but not as consolidated as the underlying 'basal peat' with Ericaceae and Cyperaceae present, together with some *Sphagnum*. In the Pennine areas, and indeed elsewhere, a third major stratigraphical division can be made. This 'upper peat' is less compacted than either the basal or the lower peats and contains abundant *Sphagnum*. This change took place in the Pennines c.1200-1500 BC, but elsewhere is often later, c.600 BC. Peat growth accelerated at this time, rates of 3.3-4.7 cm/100 yr being estimated (Tallis 1964). This level of renewed growth ('recurrence surface') in peat bogs at c.600 BC is widespread and is often referred to as the Grenzhorizont, following Weber. His hypothesis was that in the dry Sub-Boreal, the bog surface became dessicated, the peat humified and little peat deposited ('retardation layer'). Other workers attribute many so-called recurrence surfaces to local changes in drainage (Moore & Bellamy 1974). Whatever the exact cause of these humified or 'retardation layers' might be, their presence is probably important in accounting for the observed relative impermeability of peatlands.

3. Observations indicating the significance of layering in peat in fluvial processes

At a bog-burst near Killarny, at the end of the 19th century, the underlying more humified peat flowed out from beneath the upper, more fibrous layer (Sollas et al 1897). This was also believed to be the case at flows in Co. Mayo (Delap et al 1932), Co. Clare (Mitchell 1935) and in Antrim (Colhoun et al 1962). At the bog bursts at Stainmore (Huddleston 1930), in Co. Wicklow (Mitchell 1938; Delap & Micthell 1939) and in Co. Donegal (Bishopp & Mitchell 1946), there is no evidence of the lower peat moving out. In some cases the 'burst' seems to have been more of a slump, with the peat moving along a lubricated layer. Others are more turbulent flow but over a firm basal layer.

It has been assumed that '... deep blanket peat is inherently unstable ... owing to the great volume of water which saturates the peat" (Johnson & Dunham 1963) and Sollas et al (1897), in a review of all bursts then known, had discussed whether water drains through peat into the lower layers. This would cause them to become saturated and to flow out. In contrast modern observations of the hydraulic conductivity of peats suggest that such percolation is not common. Furthermore, examination of the description of bog bursts with such underflow suggests that these flows or bursts had pre-existing drainage channels, either within or under the peat (cf. Ingram 1967). (For example, the 1931 Mayo burst originated where water issued in a rising, being water which drained from a lake 500 yards upslope.) The 1930 Stainmore bursts left 3 m vertical scars in the peat "... at the northern end where the gill itself begins" (Huddleston 1930). Even those bursts which seem to occur at a break of slope often have areas of water collection behind and a stream below, as described by Colhoun et al (1965).

(It seems that the occasional, sudden surge of water released by bog bursts in headstreams is attributable to a condition where the drier peats at the bog edge act to bank the water-carrying peat in central areas of the bog. When heavy rainfall occurs, the additional weight causes a break at the point of weakness. At sites with internal drainage or drainage below the peat, the lower layers may pour out first, rafting the upper layers as they are torn away. Alternatively, if the under-peat does not have internal drainage, then slumping may occur on a lubricated lower surface.)

4. Organisation of internal drainage within peatlands

The presence and the probable hydrological significance of recurrence surfaces and retardation layers in peat profiles, together with their low hydraulic conducticity, argue against the explanation of 'sinkholes' and subsurface channels in peatlands by analogies with 'karst' drainage. In the Brishie Bog, a 'sinkhole' is located where the peat of the bog itself abuts on to the blanket peat of the sloping ground to the north (Fig 9.2). On the south side of the bog, two 'sinkholes' lie at the head of a small, shallow valley extending back from the rand. This valley corresponds with a valley in the underlying mineral ground, and it may be that these 'sinks' have developed by slumping. Runoff is not seen to enter the 'sinks'. In dry weather, with a low watertable, the sinks are dry, but in wet, high watertable conditions, the sinks appear as pools with the water gradually entering through the surface layer, not in concentrated surface flow. Water loss is a slow, gradual process. This variation in appearance with the height of the watertable suggests that these 'sinks' may not extend to the base of the peat, but that water escapes at a higher layer in the peat. Although these sinks may be cracks developing as the peat dries as a result of growth of the rand valley, to liken their hydrological role with that of master joints in limestone would be misleading.

Perhaps some first steps to comprehending the internal drainage of peatlands are better made on the assumption that its critical features could be unique to such areas, rather than by analogies with more closely studied sub-surface drainage patterns in soil or rock. This possibility is being considered in the peatlands of the Antrim Plateau, above Carnlough, where a 100-300 m wide peat bench, which is gently undulating in cross-profile, has either small

Figure 9.6   Surface drainage 'soakways' across peatland on the Antrim Plateau above Carnlough. Surface flow concentrates on the ridge from which the view is taken before flowing on to the peat-covered bench in the area to the left of the figure in the foreground. The soakway swings to the right before ending in a sinkhole, where the steep straight side of the more consolidated peat is clearly seen, to the right of the centre of the view

streams or gently inclined soakways running across it (Fig 9.6). The intervening 'highs' of peat are drier and more consolidated. A 'sinkhole' with a distinctive D-shaped plan and approximately 4 m in diameter lies near the head of this bench, some 50 m from the break of slope. A wide depression, no more than 10-20 cm below the general peat level, leads from the slopes to the sink-hole. It is more 'spongy', with much *Sphagnum papillosum* and some *Molinia, Carex spp.* and a variety of mosses. Significantly this vegetation, markedly different from the surrounding peat, continues beyond the sink towards the river. The straight side of the 'sinkhole' is the edge of the more consolidated peat in which peat growth has continued and left an underground channel beneath on the downslope side of the sink.

One possible interpretation for concentrated linear sub-surface flows in peatlands is that the headstreams, developed in mineral ground, have become overgrown by peat rather than collapsing later. This certainly happens with abandoned drainage ditches. Both natural and man-made ditches are observed to 'heal' quickly in peatlands (Mayfield & Pearson 1972). Since peatland is, in part, a living community, to describe this process as 'healing' is scarcely metaphorical. Studies of peat erosion suggest that the process is of relatively recent date. Tallis (1973) ascribes peatland gully erosion to the last 200-250 years, Thomas (1956) dates gully erosion in South Wales to the last 500 years, whilst Imeson (1971) demonstrates that gully erosion is extremely rapid following burning, the gullies extending back by the action of runnels down the peat face. Hydrological studies, such as they are, suggest that ground flow is a minor part of runoff from bogs, and peat erosion seems to have occurred in relatively recent times. Instability due to human land usages on or around the peat areas may be a more likely explanation of any contemporary gullying in peatlands, rather than its being an intrinsic tendency in the organisation of internal drainage within peat.

CONCLUSION

Clearly there are several factors affecting the runoff from peat bogs. The height of the watertable affects storage capacity and runoff. Evaporation is also affected, since increasing depth limits evaporation. Watertable height is dependent upon atmospheric factors and on hydrological properties of the mire, such as its hydraulic conductivity and gradient, both of which in turn depend on the vegetational composition of the peat. The assumed gradual release of water to streams in drier periods seems to be in question. The fact that a peatland may feel like a sponge does not imply that this analogy extends to water release and retention. As the beginning of peat growth in upland Britain is usually given as around 5500 BC, the start of the Atlantic Period or Zone VIIa, it is quite conceivable that some peatlands have merely covered over a pre-existing, integrated channel system of drainage. This possibility, together with the problems of distinguishing weather and climate changes from the effects of human usage, does not detract from the significance of peatlands as the starting points, past and present, of the 1st-order head-streams of the major rivers in cool-temperate environments.

ACKNOWLEDGEMENTS

I am indebted to Dr. D. J. Boatman, Department of Plant Biology, University of Hull, with whom the work on the Silver Flowe was carried out, to Mrs. M. M. Cruickshank (nee Bower), Department of Geography, Queen's University of Belfast, for much helpful discussion, and to the cartographers of the Department.

REFERENCES

Bishopp, D. W. & Mitchell, G.F., 1946. On a recent bog-flow in Meenacharvy Townland, Co. Donegal. *Scientific Proceedings of the Royal Dublin Society*, 24, 151-156

Boatman, D.J. & Tomlinson, R.W., 1973. The Silver Flowe I. Some structural and hydrological features of Brishie Bog and their bearing on pool formation. *Journal of Ecology*, 61, 653-666

Boelter, D.H., 1965. Hydraulic conductivity of peats. *Soil Science*, 100, 227-231

Bower, M.M., 1961. The distribution of erosion in blanket peat bogs in the Pennines. *Transactions of the Institute of British Geographers*, 29, 17-30

Bower, M.M., 1962. The cause of erosion in blanket peat lands. *Scottish Geographical Magazine*, 78, 33-43

Chapman, S.B., 1965. The ecology of Coom Rigg Moss, Northumberland. III. Some water relations of the bog system. *Journal of Ecology*, 53, 371-384

Clapham, A.R., Tutin, T.G. & Warburg, E.F., 1962. *Flora of the British Isles*. (Cambridge University Press, London), 2nd edition

Colhoun, E. et al., 1965. Recent bog flows and debris slides in the north of Ireland. *Scientific Proceedings of the Royal Dublin Society Series A*, 2, 163-174

Conway, V. & Millar, A., 1960. The hydrology of some small peat covered catchments in the northern Pennines. *Journal of the Institution of Water Engineers*, 14, 415-424

Daniels, R.E. & Pearson, M.C., 1974. Ecological studies at Roydon Common, Norfolk. *Journal of Ecology*, 62, 127-150

Delap, A.D. et al., 1932. Report on the recent bog-flow at Glencullin, Co. Mayo. *Scientific Proceedings of the Royal Dublin Society*, 20, 181-192

Delap, A.D. & Mitchell, G.F., 1939. On a recent bog-flow in Powerscourt Mountain Townland, Co. Wicklow. *Scientific Proceedings of the Royal Dublin Society*, 22, 195-198

Goode, D.A., 1970. *Ecological studies on the Silver Flowe Nature Reserve*. Unpublished PhD thesis, University of Hull

Huddleston, F., 1930. The cloudbursts on Stainmore, Westmorland, June 18, 1930. *British Rainfall*, 287-292

Imeson, A.C., 1971. Heather burning and soil erosion on the North Yorkshire Moors. *Journal of Applied Ecology*, 8, 537-542

Ingram, H.A.P., 1967. Problems of hydrology and plant distribution in mires. *Journal of Ecology*, 55, 711-724

James, P.W., 1965. A new check list of British lichens. *Lichenologist*, 3, 95-153

Johnson, G.A.L. & Dunham, K.C., 1963. The geology of Moor House. *Monographs of the Nature Conservancy*, 2, (Her Majesty's Stationery Office, London)

Mayfield, B. & Pearson, M.C., 1972. Human interference with the north Derbyshire blanket peat. *East Midland Geographer*, 5, 245-251

Mitchell, G.F., 1935. On a recent bog-flow in the Co. Clare. *Scientific Proceedings of the Royal Dublin Society*, 21, 247-251

Mitchell, G.F., 1938. On a recent bog-flow in the Co. Wicklow. *Scientific Proceedings of the Royal Dublin Society*, 22, 49-54

Moore, P.D. & Bellamy, D.J., 1974. *Peatlands*. (Elek Science, London)

Pringle, J., 1948. *British regional geology: the south of Scotland* (Her Majesty's Stationery Office, Edinburgh), 2nd edition

Ratcliffe, D.A. & Walker, D., 1958. The Silver Flowe, Galloway, Scotland. *Journal of Ecology*, 46, 407-445

Richards, P.W. & Wallace, E.C., 1950. An annotated list of British mosses. *Transactions of the British Bryological Society*, 1, 1-31

Robertson, R.A. & Jowsey, P.C., 1968. Peat resources and development in the United Kingdom. *Proceedings of the 3rd International Peat Congress, Quebec*, 13

Robertson, R.A. et al., 1968. Runoff studies on a peat catchment. *Proceedings of the 2nd International Peat Congress, Leningrad* (Her Majesty's Stationery Office, Edinburgh), 161-166

Rycroft, D.W., 1971. *On the hydrology of peat*. Unpublished PhD thesis, University of Dundee

Rycroft, D.W. et al.,1975. The transmission of water through peat. I. Review. *Journal of Ecology*, 63, 535-556. II. Field experiments. *Journal of Ecology*, 63, 557-568

Sollas, W.J. et al., 1897. Report of the Committee ... to investigate the recent bog-flow in Kerry. *Scientific Proceedings of the Royal Dublin Society*, 8, 475-507

Tallis, J.H., 1964. Studies on southern Pennine peats. II. The pattern of erosion. *Journal of Ecology*, 52, 333-344

Tallis, J.H., 1973. Studies on southern Pennine peats. V. Direct observations on peat erosion and peat hydrology at Featherbed Moss, Derbyshire. *Journal of Ecology*, 61, 1-22

Thomas, T.M., 1956. Gully erosion in the Brecon Beacons area, South Wales. *Geography*, 41, 99-107

CHAPTER 10 UNDERGROUND CONTRIBUTIONS TO SURFACE FLOW, AS ESTIMATED BY WATER TEMPERATURE VARIABILITY

A. F. Pitty

INTRODUCTION

For streams to be perennial at their sources, there must be groundwater flow at the head of the stream channel (Horton 1945). In studies based on maps or air photographs, it may be sufficiently unambiguous to state that the "... points farthest upstream in a channel network are termed *sources*" (Shreve 1966). Actually, at the head of real streams the point at which flow commences lacks the persistence of a blue line on the map. Also, the degree to which the flow originates from the ground surface, soil, peat, or rock joint may seem indecipherable. A similar problem might be considered further downstream since, even at times of maximum surface runoff, groundwater flow may represent a considerable fraction of the total flow (Horton 1945). Commonly, the groundwater component in streamflow is not distinguished since few criteria separate the flow components which generate a given hydrograph pattern. Estimates of groundwater discharge of streams, or baseflow, are usually made by assuming that all flood peaks are overland-flow discharge and that the remaining lower flow segments of the hydrograph are baseflow. Occasionally hydrologists measure solutes or tracers to follow water movements through soils and ground water aquifers, but actual observations in the field are costly and slow.

The purpose of the following account falls into three distinct phases. The first is to focus attention on some points at which surface flows actually start and to demonstrate that the study of the seasonal variability of water temperature appears to afford a simple, general, and inexpensive field method for discriminating between varying origins of flow. Since the medium through which water moves underground acts as a heat source or sink, depending on season, the basic assumption of these studies is that the deeper the origin of the water and the more slow its movement, the more constant the temperature of the emerging groundwater will be. Conversely, the greater the volume of concentrated flow and the more rapid its movement, the less it will be affected by its passage underground and the narrow temperature range of the solid medium through which it has passed. Provided that the issues of groundwater are sufficiently voluminous, the modification of water temperatures by atmospheric temperatures is scarcely perceptible at the point of issue. Indeed, it will be shown that temperature measurements on surface streams give some indication of the significance of groundwater contributions to total stream flow. The data presented reveal that water temperatures, although inevitably susceptible to rapid change, nevertheless show, in their degrees of variability over a year, a striking consistency and meaningful groupings are clearly recognisable. In a second phase of the account, some interpretations of fluvial forms and processes, past and present, are briefly reviewed since a closer consideration of the nature and significance of groundwater might be justified in some future geomorphological enquiries. Thirdly, since in either hydrological or in geomorphological studies, water temperature variability appears to be a useful index, the effectiveness of a reconnaissance, two-sample approach is evaluated.

OBSERVATIONS OF WATER TEMPERATURE VARIABILITY

The sites at which water temperature variability is reported are from three main geographical groups. First, data from 35 sites in the southern Pennines is considered, particularly from the hydrologically and geomorphologically contrasted areas around Castleton, along the eastward trending Wye valley, and down the Dove and Manifold valleys in the south-west quarter of the Peak District. Secondly, 36 sites represent north-west Yorkshire ranging from Chapel-le-Dale in the west to Wharfedale in the east. In contrast to the shale, sandstone and limestone lithologies in the Pennines, observations were also made at 21 points along the southern edge of the North Yorkshire Moors anticline of Jurassic strata between Helmsley and Scarborough. The southern Pennine sites were visited on 24 occasions, those in north-west Yorkshire 12-16 times, and the North Yorkshire Moors sites were visited on 18 occasions. Although the striking lag between groundwater temperatures and air temperatures has already been described for the Pennine sites (Pitty 1976), the standard deviation is considered here in part because calculations are much fewer, but also because longer lags and reduced variability are inter-correlated. As there is always the possibility that the water temperature variability can be slightly exaggerated by contact with the atmosphere, particularly at very low flows during dry periods in summer, data is reported as being less than the calculated standard deviation rather than equal to that figure. This possible exaggeration is attributable to critically small volumes at the risings and to thermal stratification in deeper flows.

Once the standard deviations of water temperature are grouped and presented in histograms (Fig 10.1), including the data from Dartmoor tabulated in Chapter 2 (Table 2.2), some readily discernible features emerge. It is at once striking that an environmental characteristic so susceptible to modification should, nonetheless, show such consistency in the degrees of variability measured. In the southern Pennines, the most frequently observed category is that between 0.40 and $0.80^{0}C$ s.d., made up of waters in, or issuing from, the largest caves in the area. At Castleton these include the resurgence from Peak Cavern (s.d. <0.41) and the resurgences from Russet Well (s.d. <0.53) and Slop

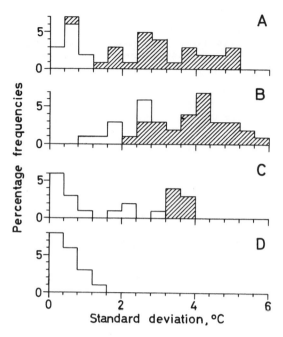

Figure 10.1 Regional comparisons between water temperature variability and between risings (blank) and surface flows (shaded).
A. Rivers and major risings in the Peak District
B. Rivers and some major risings in north-west Yorkshire
C. Rivers and major risings draining the southern flank of the North Yorkshire Moors
D. Springs in the granite catchment of the Narrator Brook (from Ternan & Williams, above, Table 2.2)

Moll (s.d. <0.68). Their combined volume of flow of 14.8 cusecs (Pitty 1968) is such that the apparently unexceptional surface stream of Peakshole Water has an unusually narrow water temperature variability which persists to the centre of the town (s.d. <0.69). Near Buxton the cave stream in Poole's Cavern (s.d. <0.78) falls in the same category, as does a stagnant pool within the cave (s.d. <0.61). At Wye Head, the starting point of one of Britain's better known rivers, the water temperature variability is s.d.<0.66. The narrowest variations include two of the main sources of water in the headstream area of the river Dove, the risings at Dowel (s.d. <0.27) and the Crowdwell (s.d. <0.31) in an area of Carboniferous Limestone in which cave development is notably absent. The least variable water temperature observed in the southern Pennines was at a point well within the Peak Cavern where an underground river emerges from a sump (s.d. <0.08). The water filling a sump is probably a significant factor here, providing an effective air-lock which seals off the inner portions of the underground drainage network from even slight atmospheric influences. A secondary mode in the Peak District observations lies between s.d. 2.4°C and 3.2°C. The 2.4-2.8°C category includes the lower points on the Wye such as Upperdale below Cressbrook (s.d. <2.42), Ashford (s.d. <2.61) and Bakewell (s.d. <2.47) and also the river Dove just before it joins the Manifold (s.d. <2.49). These relatively narrow variations are a revealing index of the degree to which these rivers are fed largely from groundwater. The 2.8-3.2°C category includes points higher up the Dove which are closer to surface runoff areas as at Crowdecote (s.d. <3.01) and at Beresford Dale (s.d. <3.14). The widest fluctuations are observed on minor streams draining off moorlands and, strikingly, the river Manifold maintains such a character until it sinks near Wettonmill (s.d. <5.00). The Dove, upstream from the major rising at Crowdecote, is similar (s.d. <5.14) but within 100 m of its junction with the Crowdwell rising which at times can double the combined flow (Pitty 1966), the overall temperature variability narrows (s.d. <3.01).

In north-west Yorkshire the modal class (Fig 10.1B) is one which contrasts with the southern Pennines pattern, being in the 4.0-4.4°C class. This category comprises the main rivers, like the Ribble at Helwith Bridge (s.d. <4.04) and the Wharfe at Coniston (s.d. <4.26) and at Bolton Bridge (s.d. <4.02) together with some minor headstreams draining from peat or drift. The secondary mode is the same as that for the Peak District in the southern Pennines, but in north-west Yorkshire the members of this group are underground rivers as, or soon after, they re-emerge. Most significant in this category is the Aire source at Airehead risings below Malham (s.d. <2.50) since Winterscales Beck below resurgences at Gatekirk (s.d.<2.42) and Brant's Gill at Horton (s.d. <2.47) were sampled at or below confluences of surface flows. Clearly the resurgence of the river Greta at God's Bridge (s.d. <1.83), that of the Alum Pot drainage at Turn Dub (s.d. <1.92) and the resurgence of Clapham Beck at Ingleborough Cavern (s.d. <1.71) are more representative of the 'underground river' category of groundwater on emergence. However, in the case of Airehead the degree of variability of the water at emergence is noteworthy since it represents the resurgence of water from Malham Tarn. The Tarn is a shallow lake which freezes over readily in winter and has the most variable of water temperatures of any of the localities studied (s.d. <5.74).

In the North Yorkshire Moors, the streams draining southwards, at least close to the points where they enter the Vale of Pickering, all fall in the 3.2-4.0°C category (Fig 10.1C). As such these rivers are intermediary between the southern Pennine rivers and those of north-west Yorkshire. In fact, their values are more closely spaced than figure 10.1C reveals. From west to east, the values run from Hodge Beck (s.d. <3.71), river Dove (s.d.<3.45), Hutton Beck (s.d. <3.53), Pickering Beck (s.d. <3.33), Thornton Beck (s.d. <3.47) to the river Derwent at the lower end of the Forge Valley (s.d. <3.77). In selecting those points at which water actually rises for sampling, there was little prior indication of the degree of variability which was revealed in the measurements. Ebberston risings (s.d. <0.10-0.12) are sufficient to provide a local water supply whilst those at Brompton (s.d. <0.11) supply the ponds which dominate the village scene, as does Keld Head (s.d. <0.04) 1 km to the northwest of Pickering. In contrast, Hutton Beck where it re-emerges in the east bank of the river Dove at Bog Hall, having passed beneath the intervening watershed, must have travelled quickly at shallow depth (s.d. <2.88). Hodge Beck resurgence (s.d. <2.12), now supplying a trout farm, has perhaps a slower underground passage since mud and silt fringe the resurgence pool compared with the sand on the floor of the Bog Hall resurgence. Since it appears to be almost stagnant, the relatively narrow variability of the pool at Harome (s.d. <2.13) may indicate a fairly deep source at this point well within the Vale of Pickering. The resurgence of the river Rye at Rye House (s.d. <1.84) is close in variability to that of the major resurgences in north-west Yorkshire. However, there are several indications that water enters and travels through the Jurassic strata by a large number of separate, slightly widened joints rather than through one major conduit.

Underground drainage in massive limestone has the distinctive characteristic of being actually observable along some parts of its underground course. Otherwise, it is simply the end member of a series of aquifer types rather than a curiosity to be set on one side, together with other karst features. Thus the summary of comparable data from Ternan & Williams' work (Chapter 2) for springs in a granite area is a striking inclusion in Figure 10.1 as the grouping of values (Fig 10.1D) matches closely that of groundwaters emerging from Jurassic strata. The major general advantage of studying groundwaters in massive limestone areas is that the "auxiliary outlets for groundwater" postulated by R. E. Horton (1936) can be readily identified and, in some cases, explored and mapped. Also, groundwater entering a stream channel at a particular point, rather than being simply seepage from bank storage, is clearly observable at dry-weather resurgences within the confines of stream channels as in the Skirfare channel above Arncliffe, the Greta at God's Bridge, or the Manifold at Ilam. In perennially flowing streams such "stream channel outlets" of groundwater might be difficult to observe directly due to turbulence of the water or sediment on the channel floor. Therefore the existence of "stream channel outlets" for groundwater is postulated, and water temperature characteristics suggested as the criteria by which they might be identified.

GEOMORPHOLOGICAL SIGNIFICANCE OF GROUNDWATER

In the study of fluvial processes, attention readily turns to occasions when the sheer repetition of the annual or bi-annual flood accounts for the removal of the bulk of the debris from a drainage basin. Groundwater contributions, if assumed to be no more than mere baseflow on the hydrograph, appear undramatic compared with the pulse of flow

Underground contributions to surface flow

which follows heavy rain and which is conventionally attributed to runoff. However, Horton (1936) pointed out that long-continued rains would not necessarily produce surface runoff nor increase in groundwater flow until groundwater levels reached that of the auxiliary outlets for groundwater flow. At such a time, rain of no great intensity would produce abruptly a relatively high discharge, substantially the same effect as if infiltration had been reduced to zero.

G. T. Warwick (1960) has considered the possibility of groundwater characteristics as an influence on stream longitudinal profiles and in the development of incised valley meanders, and several studies consider the implications of *loss* on stream water to a deeper seated groundwater body. It is, however, the possible significance of *increases* in stream flow from groundwater which might be profitably evaluated in geomorphological studies. For example, since much of the flow in the Derbyshire Wye is essentially groundwater coming to the surface, it seems possible that some associated valley forms and features might be linked with increases in groundwater flow. Water temperatures draw attention to the 5 km reach between Topley Pike (s.d. <1.97) and Millers Dale (s.d. <1.83), with the variability of water temperature diminishing downstream. Much of the inflow from groundwater occurs at the lower end of the pair of valley meanders around Chee Tor. Another pair of valley meanders are well-developed downstream, just before further groundwater contributions enter the main river at Cressbrook. In both cases, the amplitude of the valley meanders increases downstream. For instance the contour some 60 m above the stream has a radius of approximately 300 m, increasing to 450 m around Chee Tor and the pair of valley meanders above Cressbrook increase from 400 m to 500 m radius for the same contour. Over this total reach the estimated mean discharge doubles from 42 cusecs at Topley Pike to 84 cusecs just below Cressbrook (Pitty 1968). In the case of the Manifold, where surface flow decreases and usually ceases after passing on the limestone outcrop, the two impressive valley bends at Ecton show a reversed trend, with the radius of the contours 60 m above the stream decreasing from 550 m to 450 m. If these three instances are related to the pronounced changes in groundwater characteristics at these points, the increase or decrease in surface flow at these points must have been established for the appreciable time which would be required to create valley bends on this scale. Indeed, in the case of the upstream bend at Ecton, a 25 m thick bank of cemented scree, by extending down to river level, indicates that this valley bend is, at the latest, a product of the last interglacial (Prentice & Morris 1959). In a series of papers in which valley bends were investigated in great detail, G. H. Dury (1966) has concluded that valley meanders were cut at times when discharge was greater than at present, and attributed to much heavier rainfall in the past than at present. When he claimed that "... the occurrence of valley windings has been shown to be independent of movements of the water table..." G. H. Dury (1954) was considering only the possibility of diminished discharge possibly being due to a *loss* of surface flow to groundwater. That diminished present-day discharges could be due to a reduction in the occurrence of abrupt overflows from auxiliary outlets of groundwater is an alternative possibility worth considering. However, as the mechanisms by which a river can create symmetrical bends in hard rock remain an unsolved problem, no further speculations are justified here. These comments are intended merely to emphasize that striking landform characteristics occur at points where groundwater conditions change and that water temperature characteristics can draw attention to their unexplained association.

In addition to water temperatures as an index of groundwater and surface flow admixtures, the possible significance of water temperature variability in geomorphological and environmental processes might be considered briefly. The fact that the Ecton scree developed on such a striking scale compared with other limestone valley slopes in the southern Pennines might be linked with the Manifold having strictly the thermal characteristics of a surface stream (s.d. <5.00) and a mean of $8.74°C$. At the most, an episodic summer flow might have been maintained during the last glacial maxima. In the contrasted case of the river Wye at Millers Dale (s.d. <1.83), with a present-day mean of $9.03°C$, the chances seem greater of more persistent surface flow in times of colder climate. Taking $5°C$ as a general figure for estimated drop in world temperatures during a glacial maximum in conjunction with the very small variability recorded for present-day water temperature in the major risings draining on to the Vale of Pickering, a striking possibility is suggested. During a glacial advance, this zone could have remained one of a steady outflow of groundwater with a temperature departing little from one not falling much below $4°C$. Both the Peak District and the North Yorkshire Moors appear to have been enclaves and drainage diversions due to ponding of water by ice have been postulated. The present data on water temperature variability in these two areas suggests that the converse situation is at least worth considering. Rather than the glacier ice ponding surface drainage, it is conceivable that the outflow of groundwater was sufficiently steady in volume and temperature to arrest the spread of glacier ice. A glacier in the Derbyshire Wye valley is as likely as the need to complain to the waiter about finding ice-cubes in one's soup.

RECONNAISSANCE SURVEY OF WATER TEMPERATURES

It has been demonstrated that water temperature variability provides an instructive index of groundwater characteristics at stream sources, and that in surface streams baseflow contributions are detectable in a similar way. Furthermore, groundwater and its thermal characteristics themselves may have been underestimated in geomorphological process studies. To explore these possibilities further, and preferably at a larger number of points, a more rapid sampling procedure is needed than the repeated visits necessary to establish a reliable standard deviation.

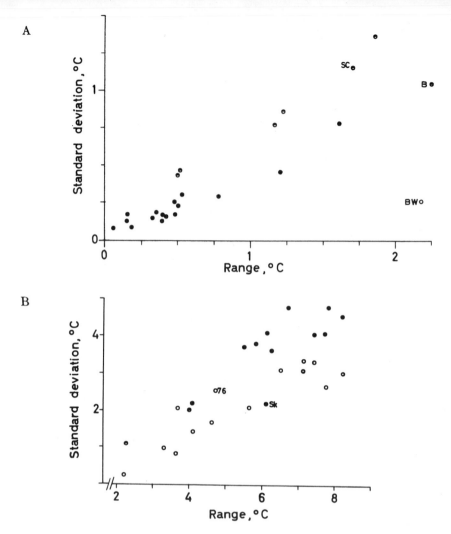

Figure 10.2   Standard deviation of water temperatures in Littondale and Wharfedale established by Ternan (1972) and by Whittel (1978) as estimated independently by a two-sample six-month range.
A. Smaller risings (full circles), including some at which the actual point of emergence is indistinct in dry weather (half-open circles). In the latter cases, such as Scoska Cave (SC), slight modification of water temperatures by air temperature occurs. Other localities indicated are the Brow Well (BW) resurgence, Grassington, and one of the risings in the Blackhill col (B), site 92 of Ternan (1972).
B. Larger risings, resurgences and surface streams. The resurgences (open circles) tend to fall below the general trend, whereas surface streams (closed circles) have comparatively larger standard deviations for a given observed range. The sites indicated, the Skirfare (Sk) upstream from Arncliffe and site 76 (Ternan 1972), are discussed in the text.

Therefore, several sites for which the standard deviation of water temperatures had already been established independently, on the basis of up to 24 observations at each site, were visited twice. First, after snowmelt in late February, and again in the late summer after rainfall in early September. This timing for the 6-month interval was due to the known lag of some weeks in the water temperature of risings behind that of air temperatures. The advantage of choosing sampling dates after snowmelt or rainfall is that the more rapid underground flows might be more clearly distinguished from sources of deeper-seated or slower-moving groundwater.

The data does, in fact, fall into two distinct groups, with two distinctive risings occurring at the transition zone where the range was about 2^0C. Although the Blackhill col rising (Fig 10.2A) emerges from a tight joint, its wide temperature variation compared with the narrowness which typifies springs, is consistent with traces of silt being deposited around its outlet. Conversely, whilst the voluminous Brow Well resurgence, close to the North Craven fault near Grassington, is normally indistinguishable from springs (s.d. <0.28), the resurgence component in its flow is revealed at the extremes of snowmelt and summer rainfalls. Figure 10.2A shows that, for risings where the seasonal range was less than 0.50^0C, the range relates to the standard deviation closely. Between 0.50^0C and 2.0^0C range, the relationship is maintained, the greater scatter being attributable to the rising water at some sources emerging within a small zone rather than at a uniquely definable point. The surface streams (Fig 10.2B) extend the general trend of the springs, with the ratio of observed range being, in general, twice the standard deviation. The major resurgences fall consistently along this general trend, whilst most of the points lying above the general trend are a sequence of sampling points on the river Wharfe (Whittel 1978). In the case of repeated sampling of surface streams, diurnal variations are probably a significant additional source of variation.

The two apparently anomalous points on Figure 10.2B merely support the general theme of present argument. The river Skirfare, just upstream from Arncliffe, appears to be a 'typical' surface stream at this point. In fact, low flow conditions leave the river bed dry for a 3 km reach upvalley to Litton and the 'surface' flow at Arncliffe is really 'groundwater at the surface'. Conversely, at site 76 (Ternan 1972) near Blishmire, flow occurs only for about a quarter of the year, during wet weather. Thus, the water emerging from the ground at this point is, in origin, essentially surface runoff from an impermeable stratum higher in the Yoredale sequence.

CONCLUSIONS

Some leading directions in fluvial geomorphology have been defined with the separate consideration of groundwater excluded. Temperature variability is suggested as an index which can, to some extent, indicate the origin of emerging groundwaters and can, further downstream, draw attention to geohydrologically distinctive localities at which geomorphologically distinctive features are developed. Where water temperatures of emerging groundwaters vary little, the possible implications for the maintenance of local enclaves and ecological refuges during cooler climates might be considered. Although the standard deviation would be particularly useful in assessing such possibilities, a two-sample 6-month range of water temperatures is seen to be a useful first step in implementing enquiries into water temperature characteristics of emerging groundwaters.

REFERENCES

Dury, G.H., 1954. Contribution to a general theory of meandering valleys. *American Journal of Science*, 252, 193-244

Dury, G.H., 1966. Incised valley meanders on the Lower Colo, New South Wales. *The Australian Geographer*, 10, 17-25

Horton, R.E., 1936. Maximum ground-water levels. *Transactions of the American Geophysical Union*, 17, 344-357

Horton, R.E., 1945. Erosional development of streams and their drainage basins: hydrophysical approach to quantitative morphology. *Bulletin of the Geological Society of America*, 56, 275-370

Pitty, A.F., 1966. The estimation of discharge from a karst rising by natural salt dilution. *Journal of Hydrology*, 4, 63-69

Pitty, A.F., 1968. The scale and significance of solutional loss from the limestone tract of the southern Pennines. *Proceedings of the Geologists' Association*, 79, 153-177

Pitty, A.F., 1976. Water temperatures in the limestone areas of the central and southern Pennines. *Proceedings of the Yorkshire Geological Society*, 40, 601-612

Pitty, A.F. et al., 1979. The range of water temperature fluctuations in the limestone waters of the central and southern Pennines. *Journal of Hydrology, in press*

Prentice, J.E. & Morris, P.G., 1959. Cemented screes in the Manifold valley, North Staffordshire. *East Midland Geographer*, 2(11), 16-19

Shreve, R.L., 1966. Statistical law of stream numbers. *Journal of Geology*, 74, 17-37

Ternan, J.L., 1972. *Karst water studies in the Malham area, north of the Craven Faults.* Unpublished PhD thesis, University of Hull

Warwick, G.T., 1960. The effect of knick-point recession on the water-table and associated features in limestone regions, with special reference to England and Wales. *Zeitschrift für Geomorphologie*, Supplementband 2, 92-99

Whittel, P.A., 1978. *Some aspects of the fluvial geomorphology of Upper Wharfedale.* Unpublished PhD thesis, University of Hull

CHAPTER 11 DRAINAGE-NET CONTROL OF

SEDIMENTARY PARAMETERS IN

SAND-BED EPHEMERAL STREAMS

Lynne E. Frostick & I. Reid[*]

INTRODUCTION

Downstream changes in the character of river alluvium reflect a complex interaction of hydrodynamic processes whose individual importance may not be readily discernible. Where stream-bed conditions are artificially simplified and particle populations sufficiently uniform, the mechanics of individual particle movement can be successfully modelled (Francis 1973; Meland & Norrman 1966). In natural streams, the prediction of particle movement is often frustrated by complex bed geometries. At the smallest scale, particles may be more or less hidden in a bed microtopography created by larger neighbouring particles, a factor highlighted by Laronne & Carson (1976) and built into at least one of the bedload transport equations (Einstein 1950). At larger scale, and especially in sand-bed rivers, the mutual interference of mobile bedforms and fluid motion produces an unstable hydraulic geometry (Martinec 1967; Culbertson et al 1971) which accentuates differences in particle velocity according to the size, shape and density of individual grains. At still larger scale, river channel plan geometry, in directing the fluid stream, ensures a distinct pattern of shear force at the bed that changes in magnitude and configuration as discharge fluctuates (Francis & Asfari 1971; Smith 1974; Hooke 1975; Bridge & Jarvis 1976; Bridge 1977; Bathurst et al 1977). Particles may be stored more or less indefinitely in channel bars, waiting to be moved spasmodically from one position of temporary rest to another as floods of different recurrence-interval sweep the channel.

The combined result is an intricate transport pattern, with the stream a complicated and elongate mosaic of processes moving individual particles in a manner that defies generalisation. Despite this, systematic downstream changes in sedimentary parameters have been evaluated for a considerable number of streams. The majority of case-studies deal with coarse-grained alluvium though a small body of literature is pertinent to fine-grained sediment. For convenient reference a tabular review is presented (Table 11.1). The objective of most studies has been verification of a fundamental precept of sedimentology -

---

[*] Departments of Geology and Geography, Birkbeck College, University of London

Table 11.1   Downstream petrographical changes in alluvium

Author	River or System	Transport-Distance (km)	Petrographic Parameter (i increase, d decrease, n/s no significant change) Size (range in mm)	Roundness	Shape/Sphericity /Form	Petrographical Change
Sternberg, 1875, Z.Barwessen, 25	Rhine, Germany	260	d 160 → 90			
Hochenberger, 1886,	Mur, Austria	121	d 86 → 37			
Wentworth, 1922, Bull.U.S.Geol.Surv., 730-C	Russell Fork, Big Sandy R., Va.	52		i		d quartzite
Mississippi River Comm., 1935, Pap.U.S. Wat.Exp.Sta., 17	Mississippi	1770	d 0.7 → 0.2			
Russell and Taylor, 1937, J.Geol., 45	Mississippi	1770		d		d feldspar
Sidwell, 1941, J.sedim.Petrol., 11	Pecos R., N.Mex.	321				i epidote, tourmaline haematite d magnetite, ilmenite
Krumbein, 1942, Bull.Geol.Soc.Am., 53	Arroyo Seco., Calif.	18	n/s	i	n/s	d granodiorite
Plumley, 1948, J.Geol., 56	Cheyenne R., S.Dak.	64	d 56 → 4	i	i	i chert
Blissenbach, 1952, J.sedim.Petrol., 22	Alluv. fans, Ariz.	7	d 4000 → 2500			d feldspar

Table 11.1 (continued)

Author	River or System	Transport Distance (km)	Petrographic Parameter (i increase, d decrease, n/s no significant change)			
			Size (range in mm)	Roundness	Shape/Sphericity /Form	Petrographical Change
Potter, 1955, J.Geol., 63	Lafayette Gravel	103	d 12 → 0.7	n/s		
Hack, 1957, Prof.Pap.U.S.Geol.Surv., 294B	Various Appalachian Streams	16 to 113	d 600 → 210 130 → 16 90 → 37 i 7 → 80 n/s			d sandst.,limest.
Sneed and Folk, 1958, J.Geol., 66	Colorado R., Texas	241		i quartz limest. n/s chert	i quartz <38mm n/s quartz >54mm	d granite, gneiss, schist, sandst., limest. i quartz, chert
Inderbitzen, 1959, J.sedim.Petrol., 29	Alameda Cr., Calif.	29	d 100 → 25	n/s	n/s	
Brush, 1961, Prof.Pap.U.S.Geol.Surv., 282F	16 Pennsylvanian Streams	18 to 56	d 180 → 22 110 → 12 43 → 38 i 35 → 60 75 → 80 n/s			
McCammon, 1961, J.sedim.Petrol., 31	Wabash R., Ind.	402				i chert n/s quartzite

Table 11.1 (Continued)

Author	River or System	Transport-Distance (km)	Size (range in mm)	Roundness	Shape/Sphericity/Form	Petrographical Change
Pollack, 1961, J.sedim.Petrol., 31	S.Canadian Rv., N.Mex. Tex., Okl.	1046	n/s	i sphene, leucox.-ilmen. haemat. pyrox.-amphib. tourm. d quartz	i quartz feldspar haematite magnetite	n/s heavy minerals
Bluck, 1964, J.sedim.Petrol., 34	Alluv. fan, Nev.	5	d 1250 → 250		n/s	
Naidu, 1966	Godavari R., India	84	d 2 → 0.25			
Flores, 1967, J.sedim.Petrol., 37	Lower Freeport Sandst., Ohio	137				n/s quartz, feldspar mica
Ouma, 1967, Pub.Int.Ass.Sci.Hyd., 75	Hacking R., N.S.W.	29		i → d		
Scott, 1967, Pub.Int.Ass.Sci.Hyd., 75	Rubicon R., Calif.	3	d 256 → 84	i		
Dal Cin, 1968, J.sedim.Petrol., 38	Piave R., Italy	179		i quartz		i quartz d metamorphics, granite
Brattacharya, 1969, Science and Culture, 35	Ajay R., India	–		d		

Table 11.1 (Continued)

Author	River or System	Transport-Distance (km)	Petrographic Parameter (i increase, d decrease, n/s no significant change)			
			Size (range in mm)	Roundness	Shape/Sphericity/Form	Petrographical Change
Whetten et al, 1969, J.sedim.Petrol., 39	Columbia R., Can.	900	i 0.004 → 2			i feldspar, rock fragments
Bradley, 1970, Bull.Geol.Soc.Am., 81	Colorado R., Tex.	386	d 135 → 44 chert 105 → 28 quartz 100 → 20 granite			i graphic granite, pegmatite d aplite, gneiss, granite
Miall, 1970, J.sedim.Petrol., 40	Peel Sound, Fm., Canada	—	d 69 → 28	d	n/s	
Smith, 1970, Bull.Geol.Soc.Am., 81	Platte R., Col., Nebr.	904	d 3 → 0.3			
Cameron and Blatt, 1971, J.sedim.Petrol., 41	Elk Cr., S.Dak.	193		i volc. rock frags.		i feldspar d schist
Douglas et al, 1971,	Wadi Kuf, Libya	80	i (0-67 km) 40 → 82 d (67-80km) 82 → 60			
Pearce, 1971, J.sedim.Petrol., 41	Vicary Cr., Alberta	—		i		
Gustavson, 1974, J.sedim.Petrol., 44	Fountain, Alder Streams, Alaska	5 and 2.4	d 460 → 90 670 → 150	i		
Gregory and Walling, 1973	R. Ure, England	—	d 63 → 43	i		

Table 11.1 (Continued)

Author	River or System	Transport-Distance (km)	Petrographic Parameter (i increase, d decrease, n/s no significant change)			
			Size (range in mm) †	Roundness	Shape/Sphericity /Form	Petrographical Change
Smith, 1974, J.Geol., 82	Upper Kicking Horse R., B.Columbia	7	d 45 → 6			
Goede, 1975, J.sedim.Petrol., 45	Tambo R., Victoria	109		i sands.	n/s	
Self, 1975, J.sedim.Petrol., 45	R. Nautla, Mexico	150	d 32 → 0.5			i feldspar, limest.,rock fragments d quartz, heavy minerals
Shukis and Ethridge, 1975, J.sedim.Petrol., 45	St. Francis R., Miss.	51	d 1 → 0.13 quartz i 0.25 → 1 rock fragments			
Laronne and Carson, 1976, Sedimentology, 23	Seales Br., Quebec	0.7	d 160 → 100			
Spalletti, 1976, J.sedim.Petrol., 46	Sarmiento R., Argentina	50			i	
Frostick and Reid, this publication	Ephemeral St., Kenya	2.2	i 0.3 → 0.55			

† in some cases an average of 3 axial diameters, in others the intermediate axis

that increasing transport distance brings about an increase in sedimentary sorting, either through the elimination or comminution of particles whose mineralogy confers upon them poor mechanical durability, or through the hydraulic separation of larger or smaller, spherical or non-spherical particles.

Besides these petrographical analyses, a few studies have recorded systematic downstream changes in the morphological and structural characteristics of channel sediments. Smith (1970) shows decreasing values of bed rugosity through use of a bed-relief index. Frostick & Reid (1977) demonstrate that tributary water and sediment contribution influences the primary structures of ephemeral channel-fill.

The difficulties encountered in discerning systematic linear changes in sediment character within the stream channel because of the complex pattern of particle storage in channel bars or within bed microtopography are reduced as a stream disgorges to a large water body, such as a lake. Here, the hydrodynamics of sediment disposal become simplified (Jopling 1964; Wright & Coleman 1974). In the absence of density currents, sediment dispersal depends largely upon particle fall velocity, and offshore fining is a well-established delta characteristic (Shepard 1956; Allen 1965; Müller 1966). Subsequent littoral reworking by wave action produces a sedimentary complex of beach bars and lagoons. However, this is only local overprinting, an important but restricted complication of a general pattern established by fluvial sorting. This systematic spatial ordering of both fluvial and deltaic sedimentary facies is well-established for streams with perennial flow, but sorting patterns brought about by the peculiar nature of arid-zone flood-hydrographs are not well documented (2 references, Table 11.1).

The Koobi Fora Sedimentary Basin in northern Kenya, the location of this study (Fig 11.1), provides ideal conditions for evaluation of such ephemeral stream deposits, particularly because they are as yet unaffected by human intervention. Those of Area 103 (Koobi Fora Research Project geographical designation) drain directly to Lake Turkana, conveniently facilitating an examination of sorting processes not only within the drainage network, but also at and beyond the lake margin. The ephemeral nature of runoff together with the scour-fill processes that operate in these sand-bed channels allows the identification of alluvium attributable to single, discrete flood events, a circumstance unusual in studies of perennial rivers.

The Koobi Fora catchments offer a useful, latitudinally distinct comparison with the drainage networks of the arid south-west USA where so many principles of fluvial geomorphology have been tested (Leopold & Miller 1956; Leopold et al 1966). In both areas, emphemeral sand-bed channels dominate the landscape. However, a primary objective of the Koobi Fora study has been to establish **spatial** patterns of sedimentation *throughout* the fluvial system in order to facilitate deduction of basinwide

Sand-bed ephemeral streams

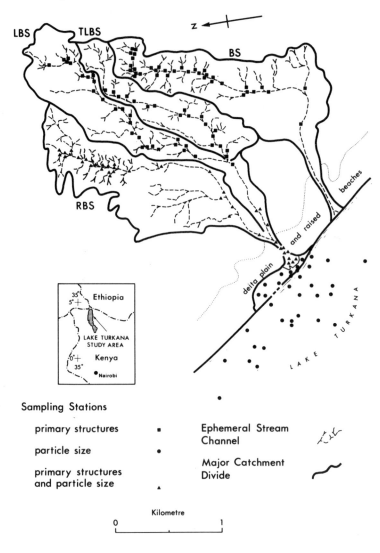

Figure 11.1 Regional location of the Lake Turkana study area and the position of sampling stations within the study area

processes that can rarely be measured directly. This provides the stratigrapher with a geographical tool for paleoenvironmental reconstruction at a finer level of resolution.

THE EXPERIMENTAL AREA

1. Regional setting

   The Koobi Fora Sedimentary Basin lies east of Lake Turkana (see inset map, Fig 11.1), an area celebrated for recent discoveries of numerous specimens of fossil man (Leakey 1970, 1976). The area is thorn-scrub desert,

Figure 11.2 Above - Headwater tributary of RBS cut into silt-clays of the Koobi Fora Formation.
Below - The dissected relief of Area 103 with the light-coloured shelly lacustrine silts of the Galana Boi (foreground) and the darker Koobi Fora Formation (middle distance). Lake Turkana can be seen in the distance.

populated by the nomadic Dasenach. Average annual rainfall at Ileret Police Post is 308 mm with two seasonal maxima. Eye-witness accounts report cellular storms following discrete paths in a manner not dissimilar to those described by Diskin & Lane (1972) and Sharon (1972) for other arid zones. High rainfall intensities and negligible interception capacities combine to produce a badlands topography, with soil pillars as ample evidence of rainsplash and surface wash, and numerous collapsed interflow pipes an indication of the rapid evacuation of infiltrating water. The consequence of such efficient water disposal is a high density of sand-bed ephemeral channels. These incise the Plio-Pleistocene sediments of the Koobi Fora Basin (Fig 11.2).

Catchments of Area 103, the study area, head in high lake-stage shelly-silts of the Galana Boi and pass on to cut a modest relief of 50 m in the Upper Member of the Koobi Fora Formation, a complex series of sands, silts and clays interbedded with sheets of algal stromatolites and tuffs (Vondra & Bowen 1976; Findlater 1976). The unconsolidated nature of the clayey-silts encourages stream incision (Fig 11.2), though this is directed by the structure, a dip of 12 degrees ensuring the development of elongate drainage networks through strike-etching (Fig 11.1). Valley sideslopes tend to be short and are typically either bare planar surfaces of indurated sandstone, or a mobile colluvium of clay-silt peds.

Close to the lake shore, stream channels cross a marginal plain of deltaic and longshore drifted sediment (Fig 11.1). This is ribbed with raised beach ridges which mark longer-term changes in lake levels (Butzer 1971). An annual rise and fall of about 1 metre (Butzer 1971) is superimposed on these gross changes and governed by seasonal fluctuation in the discharge of the River Omo, the lake's major water supply. This annual oscillation is of sedimentological importance since littoral deposits are continuously reworked as the shoreline advances and retreats.

2. Channel character

Four contiguous catchments of similar size were selected (Fig 11.1). All have a comparable topography and drainage pattern as a result of the denudation of a common lithology. In upper and middle reaches, the channel extends the entire width of the valley bottom, its sand-bed directly abutting the sloping valley sides with little thalweg incision and no terracing. This is also characteristic of lower reaches, though here the wider valley floor is punctuated by point-bars. Flood debris attached to acacia trees records a maximum water stage in excess of 1 metre. The fact that floodwaters are constrained only by the valley sideslopes results in low water-depth to -width ratios. This has important consequences for sediment entrainment.

The sand-bed is typically planar (Fig 11.3), though locally more complex bedforms interrupt the surface (Fig 11.3) Excavation reveals cut and fill of up to 2 m in lower reaches but generally less upstream where bedrock may be no more than

Figure 11.3 Above - planar lamination immediately following the passage of a flood.

Below - complex pattern of bedforms in the ephemeral Tulu Bor after a flood

6 cm below surface. Successive flood deposits are usually nested. The most recent, sitting unconformably on a scour surface that truncates previous fill, is readily identified since it lacks significant cohesion and is lighter in colour. Older material is distinguished by a degree of partial induration (the sediments abound in free calcium, Fitch et al 1975) and by a progressive colour change associated with the translocation of iron typical of arid environments.

Simplicity of longitudinal profile was regarded as an important criterion when choosing the four catchments. Each is shallowly concave. Steep headwater gradients of 0.02 give way downstream to more consistent values that range around 0.01. Local variability of channel slope is often a product of discordance of the drainage line with the underlying structure. In its middle reaches, LBS cuts two scarps which successively reduce channel gradient to 0.008 from strike values of 0.012 (Fig 11.4). There are, however, no abrupt interruptions on any of the four longitudinal profiles that would generate waterfalls, a factor that might unduly complicate the sediment sorting process.

3. Rainfall-runoff response

Very little is known about stream hydrograph generation in arid zones for the simple reason that few streams are gauged. Installation costs are rarely warranted where runoff occurs so infrequently. Nevertheless, Wooding (1966), Renard (1970) and Thornes (1976) report hydrographs that confirm eye-witness accounts of streamflow at Koobi Fora. Stage rise to peak discharge is rapid. Observers report a bore of water advancing on a dry bed (Fig 11.5). Leopold & Miller (1956) describe a similar phenomenon for New Mexico arroyos. This quick concentration of water in stream channels by overland flow and by shallow subsurface stormflow is terminated with equal abruptness as rainfall stops, with the result that the recession limb of the hydrograph is steep. Runoff ceases within a few hours.

The flood may be punctuated by discharge pulses. Renard & Keppel (1966) consider these to be a response to variations in rainfall intensity. Leopold & Miller (1956) ascribe them to the influence of channel geometry and channel storage upon the floodwave. Frostick & Reid (1977) deduce from sedimentary evidence that each tributary delivers a pulse of water to the main channel, a process whose effect upon the hydrograph is accentuated should the storm be travelling up-catchment, each tributary sub-catchment awaiting its turn to participate in water evacuation.

One factor of vital significance to arid-zone channel flow is downstream attenuation of the flood-wave by influent seepage losses to both bed and banks. The cut and fill material typical of most ephemeral channels provides a receptive storage facility for such transmission losses. Renard & Keppel (1966) report a 63 per cent reduction in peak discharge over a channel length of 12 km for one particular flood in Walnut Gulch, Arizona. At Koobi Fora, an

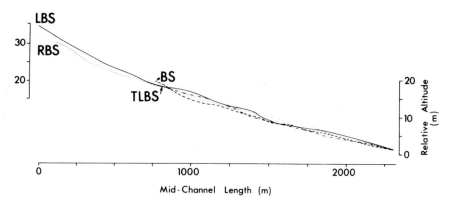

Figure 11.4 Longitudinal channel profiles

out-of-season flash-flood on the Naitiwa River was traced over less than 5 km before the bed absorbed the last vestige of water. These are comparatively large rivers. The streams of Area 103 are much smaller. While they undoubtedly suffer transmission losses, the channel-fill is not unduly deep in their upper and middle reaches, and provides only a limited storage capacity.

SAMPLING DESIGN

The channel fill of RBS above and below its confluence with LBS and of the immediate pre-confluence reach of LBS, together with their subaerial and submerged delta, beach bar and lagoon were mapped and sampled in 1974. BS, LBS and its major left-bank tributary TLBS were investigated in 1976. Subaerial sampling stations were located on aerial photographs except where the streams leave the confinement of their valleys to cross the beach-ridged marginal plain of unconsolidated material. Here channel migration, which post-dates aerial photography, precludes accurate site location and large-scale mapping becomes essential. Shore-line and off-shore sample stations were also located trigonometrically.

1. Particle-size fractionation

   Channel and sub-aerial delta samples were taken from the side-walls of pits dug specifically for the purpose. Sites were selected in the main channel of RBS to establish the effect of each tributary confluence - one sample upstream acting as local control and a second sufficiently far downstream for complete mixing to have occurred at flood discharge. Tributaries were also sampled as close as possible to their confluence with the main channel (Fig 11.1). Pits were excavated to bedrock, many as deep as 2 metres, but the results reported here concern only the most recent flood deposit. This ranges in thickness, apparently unsystematically, between 4 and 43 cm. Samples collected at each site represent all the laminae of a single flood deposit.

Sand-bed ephemeral streams

The shore-line complex of beach bar and lagoon is transient in nature, reforming each year as the lake encroaches or recedes. Bar processes are dominated by swash and backwash while the lagoon is a still-water trap for fine sediment. Unlike the channel samples, which preserve intact a single depositional episode, the beach bar and lagoonal sediments are undergoing continuous modification. Consequently surface skim samples were obtained with the intention of illustrating a pattern of sedimentary environments rather than giving them a fixed geography. Particle-size fractionation for both these and the stream channel samples was conducted at half-phi intervals using standard sieve and sedimentation procedures.

Off-shore samples were collected using a bucket dredge and dinghy. Analysis has been confined to one parameter only - maximum particle-size. Off-shore bottom contours were established by sounding with leadline close to the shore and by echo-sounder in deeper water.

2. Primary structures

Heavy- and light-minerals are segregated in alternating horizontal parallel laminae in the channel-fill throughout the four drainage networks. This has been shown (Frostick & Reid 1977) to be a useful indication of depositional mechanics during flood recession in ephemeral stream-channels. The study of RBS primary structures was extended in 1976 to LBS, TLBS and BS. As in the case of RBS, main channel sample pits were located at sufficient distance downstream from each tributary confluence to have allowed mixing at flood stage. The pit sidewalls were inspected for horizontal parallel laminae and the vertical sequence measured to the nearest millimetre.

SEDIMENTARY SORTING - PARTICLE SIZE

1. Stream channel-fill

Leaving aside for the moment the mechanism that segregates light- and heavy-mineralled particles in separate laminae, a broad view of the RBS channel sediment reveals a linear sorting from headwaters to lower reaches (Fig 11.6 A,B). Mean particle-size increases, almost doubling from 0.30 to 0.55 mm over a channel length of 2.2 km. This contrasts clearly with the size/transport relationship established for perennial streams.

The pattern of size sorting commonly expected is one of downstream fining. Although this is best illustrated for coarse alluvium, as in the classical studies of Plumley (1948) and Potter (1955), several workers have shown a similar pattern of progressive diminution in particle-size for sand-bed streams (Mississippi River Commission 1935; Pollack 1961; Naidu 1966). Basic to the precept is an assumption that (allowing for shape differences) the transport velocity of a grain is proportional to its submerged mass. Larger particles not only spend a greater portion of time in a position of rest on the stream bed, but their bigger fall velocities encourage a travel path nearer the

Figure 11.5 Above - typical bore of water moving down-channel from left to right (photo courtesy R. Watkins).
Below - large sand-bed ephemeral channel. Mature acacias characteristically line the water course relying upon perched groundwater in the sand bed.

bed where fluid velocities are reduced by boundary friction. Vertical sampling in larger rivers such as the Mississippi does indeed reveal an inverse relationship between mixing height and particle-size (Colby 1963), while Kennedy & Kouba (1970) using fluorescent sand tracer in Clear Creek, Colorado, show that particle velocity does vary inversely with the square of particle diameter. The corollary is a progressive fining of sediment with transport distance.

The Koobi Fora channel-fill appears to contradict this general principle. Such complicating factors as progressive dilution by tributary sediments, acknowledged by Whetton et al (1969) as being important for the Columbia River, can be discounted since no systematic changes in tributary sediment particle-size are detectable. In fact the downstream coarsening of the RBS bed material reflects a concomitant change in stream power. This is a product of increasing discharge, itself a function of the direct relationship between channel length and contributing drainage area (Hack 1957). Small headwater discharges, spread as they are in channels confined only by valley sideslopes, possess a power incapable of transporting anything but small particles. The successive downstream addition of tributary waters leads to progressive magnification of the flood-wave, which provides increasing stream power and ensures the entrainment of larger and larger particles. At each point the stream is more than competent to transport the finer material introduced from up-channel, and this is passed downstream leaving behind a mobile lag of appropriately coarser particles, the submerged mass of which more closely approximates the local traction threshold.

The peculiar behaviour of the ephemeral stream hydrograph now favours preservation of this leap-frog sorting pattern. In the absence of both base-flow and delayed interflow, discharge falls abruptly. The winnowing processes that resort fine particles and that are either associated with the intermediate discharges of prolonged flood recession or with the low-flow processes of perennial rivers (Crickmore 1967) are, of course, absent in arid-zone ephemeral streams. Within hours the channel is waterless and the flood deposit abandoned almost intact.

Channel deposits at A and B in the middle reaches of RBS (Fig 11.6C) provide a significant test of the sorting process. Both stations defy the downstream trend in particle-size because of unusual and localised changes in hydraulic geometry. At A, the channel is constricted (Fig 11.6C). Water passing into this reach compensates for reduction in channel width by positive adjustments in water depth and velocity, both of which increase shear forces at the bed. As a result, particles at station A are larger than the system trend-line would allow (Fig 11.6B). At station B the opposite adjustments are made since here the channel widens. The increase in hydraulic radius and decrease in velocity that result from lateral spreading of the flood water combine to reduce boundary shear so that finer particles than expected are accumulated

Channel geometry is less aberrant below B, and the trend in mean sediment size is reestablished (Fig 11.6B).

While downstream changes in particle-size are accredited mainly to the increase in stream power that accompanies increasing discharge with distance from headwaters, other variables play their part in locally bringing about deviation from a general pattern of sorting. The consequences of changing channel width have already been noted for specific sites in the middle reaches of RBS (Fig 11.6 B,C). Other pertinent variables for which data is available are channel slope and tributary sediment size. Channel width is expressed as the ratio of values downstream to upstream at each site in order to characterise local changes in hydraulic geometry. With main channel mean particle-size as the dependent variable and downstream distance, channel-width ratio, slope and tributary sediment size independent, all 5 variables were subject to partial correlation. Surprisingly the co-efficients suggest an insignificant role for channel slope and tributary sediment character, both accounting for very little of the unapportioned variance. They were, therefore, removed in subsequent analyses. The remaining variables - main channel mean sediment size (dependent), distance from headwater (distribution normalised by square-root transformation), and channel width ratio - give a multiple correlation coefficient of 0.876, significant at the 0.3 per cent level and explaining 77 per cent of the variance.

The peculiarities of the down-channel sorting process operating in RBS are reflected in the changing shape of the sediment size frequency-distribution. Transmission of fines down-channel and their imperfect evacuation from lower reaches decreases the degree of sorting in this direction (Fig 11.7A). Besides this, the distribution takes on an increasingly positive skew (Fig 11.7B). The negative values of headwater reaches reflect a comparatively high proportion of heavy-minerals whose diminutive size compensates higher specific gravities in maintaining hydraulic equivalence. Down-channel the mixing of material derived from upstream with local coarser particles, some of which are of gravel size, favours a frequency distribution with positive skew values.

2. Off-shore sediments

As flood-waters debouch into the lake, momentum is checked not only by lateral spreading of the fluid stream, but by the inertial forces of the lake water. Stream competence declines dramatically, and dispersal of the transport load depends very largely upon the fall-velocity of individual grains. The spatial pattern of particle-size established by these initial distributive processes is an off-shore fining that radiates from the point of injection. In this respect, the RBS/LBS catchment provides a typical, if small example of deltaic sedimentation (Fig 11.8). Unlike the stream channel where deposits of a single flood event are readily identifiable as one of a group of nested scour-

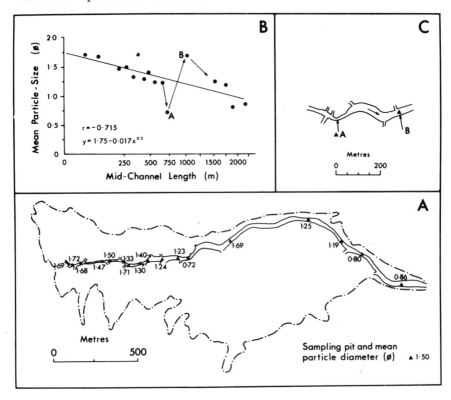

Figure 11.6 Catchment RBS channel-fill mean particle size (phi units). A. Sample site location and mean particle size. B. Mean particle size as a function of distance from headwaters. C. Channel plan configuration at sample sites A and B.

fill sets, the history of the off-shore sediments is not as easily defined. Moreover, dredge sampling does not help to unravel the sedimentary sequence. Nevertheless, by analyzing for maximum particle-size, the pattern that emerges (Fig 11.8) may be presumed independent of temporal variability in sediment supply, recording as it does the degree of spatial sorting produced by the largest recent flood discharge.

Geomorphologically the RBS/LBS delta is a minor protruberance in the off-shore contours (Fig 11.8). The frequency of flood events is probably considerably less than once a year, though no record is available, so that little material is provided for delta building. But the dramatic nature of delivery ensures coarser particles than are associated with perenially fed deltas (cf. Shepard 1956; Mathews & Shepard 1962; Allen 1965; Müller 1966; Axelsson 1967; Normark & Dickson 1976) and the particle-size gradient at the delta front is consequently much steeper.

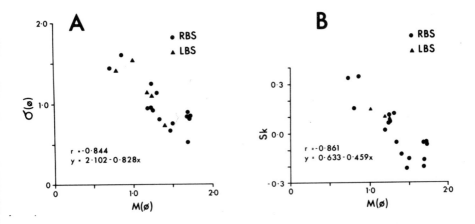

Figure 11.7 Catchment RBS channel fill graphical particle-size distribution parameters. M = mean, σ = standard deviation, Sk = skewness

3. Shore-line sediments

As the shoreline advances and retreats in response to the annual fluctuations in lake level, a limited amount of sedimentary re-working takes place. Shore processes are primarily associated with wind-waves, but wave generation on Lake Turkana is complicated by a regular diurnal pattern of changing wind direction and strength. The strongest winds come from the south-east, and therefore leave the Koobi Fora shoreline to windward. The afternoon north-westerly air stream is on-shore, but usually no more than a light breeze. As a result, wave amplitudes of about 20-30 cm are normal, though occasionally unusually strong north-westerly winds whip up waves in excess of 1 metre.

The geomorphological consequence is a small beach bar behind which lies a comparatively still-water lagoon. A schematic cross-section of the bar is given in Fig 11.9B together with a typical size-sorting pattern. The coarsest material accumulates at the wave breakpoint (mean particle size 1.3 phi or 0.41 mm) in a manner similar to that evaluated by Evans (1939), Fox et al (1966) and Kolmer (1973). The swash sweeps finer particles up on to the bar. But since a majority of waves only just overtop the bar crest, the finest particles alone (mean size 2 phi, 0.25 mm) are carried onto the planar backslope. Accelerating down this backslope, the overspill changes from sheetflow to a set of rills, the greater erosional competence of which carries whatever large particles are available towards the lagoon edge. As a result the surface sediments of the bar show a cross-sectional symmetry of particle-size (Fig 11.9B).

As long as the beach bar remains intact, it protects the lagoonal sediments from wave-action. These are the finest deposits of the drainage network, consisting mainly of a silty-clay (Fig 11.9A) with an admixture of small fossil gastropods and bivalves derived from the Koobi Fora

Sand-bed ephemeral streams

Formation. Observed flood-waters are creamy brown in colour as a result of the high concentration of suspended solids, and though many of these fine particles are distributed lakewards, a significant portion settle at the lake margin, carried there by the large-scale eddies that flank the debouchment. Some of these lagoonal sediments onto which the beach bar migrates are preserved, forming useful marker horizons in the deposits of the marginal delta plain.

SEDIMENTARY SORTING - PRIMARY STRUCTURES

Throughout the Koobi Fora drainage network the channel-fill is constructed from an alternating sequence of light- and heavy-mineral dominated horizontal parallel laminae. Only occasionally do other, geometrically more complex, primary structures interrupt the sediments. While the mechanisms that produce cross-bedding have been exhaustively studied (Allen 1963), there is little general agreement about the origin of horizontal lamination. Its very abundance (McKee et al 1967) makes it unremarkable, while its apparent dearth of parameters both measurable and useful in stratigraphical interpretation makes it the poor relation of fluvial primary structures. Pettijohn (1957, p 159) considers the alternation of coarse and fine laminae as the product of momentary fluctuations in fluid velocity, a view shared by Sanders (1965) and, in more sophisticated form, by Bridge (1978). Harms & Fahnestock (1965) and Picard & High (1973) think it indicative of plane-bed phases of sand-bed sediment transport, though no specific mechanism of accretion is defined. Jopling (1964), Smith (1971) and McBride et al (1975) all place importance on the separation of particles by size as dunes migrate along the stream bed. Under essentially steady flow conditions, coarser material is trapped in the lee of the bedform and overriden by the preponderantly finer material of the dune. Moss (1963) and Kuenen (1966) propose a radically different mechanism. The bed becomes dominated by sedimented congregations of like-particles which reject grains of differing hydraulic behaviour. For instance, a particle larger than those of the congregation protrudes higher into the fluid stream and is subject to greater torque, so increasing its chances of re-entrainment. Should the traction carpet be exhausted of any particular particle-size through continuous deposition, then a different grain-size becomes dominant.

The 1974 channel excavations at Koobi Fora revealed a systematic increase in the number of laminae downstream from catchment headwaters, a discovery arising almost entirely from the application of a geographical approach to fluvial sedimentation. In all vertical sequences a coarse-grained quartz-dominated layer is succeeded by a layer of finer grains consisting preponderantly of heavy-minerals. The number of such paired laminae corresponds closely to the number of tributary confluences above any point on the main water-course (Frostick & Reid 1977). This attribute of the Koobi Fora channel-fill points to an overriding hydrological control of horizontal lamination.

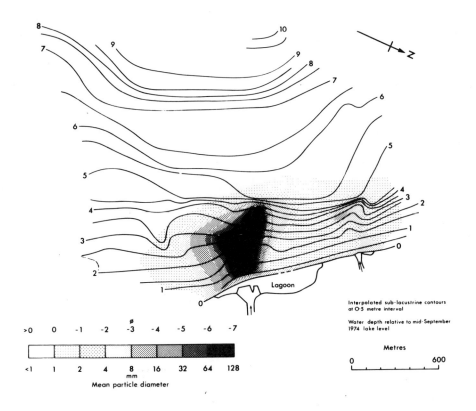

Figure 11.8   Sub-lacustrine contours and maximum particle size of the RBS/LBS delta

The shape of the flood-wave is modified by the direction of storm-travel (Yen & Chow 1969; Roberts & Klingeman 1970; Black 1972) particularly where the drainage network is as elongate as in the case of these four Koobi Fora basins (Ward 1975). Rain cells moving downchannel encourage compact hydrographs since the greater gathering-time of more distant headwaters is compensated by an earlier onset of rainfall. Up-catchment storm-travel 'stretches' the flood-hydrograph - headwater tributaries may be awaiting their turn to participate in a runoff process already under way in those closer to the catchment outfall. Storm movement transverse to the drainage network produces a hydrograph of intermediate shape, but, as in the case of up-catchment storm-travel, tributary contributions to the main channel remain more or less distinct, taking the form of a succession of sediment-laden discharge pulses. The number of such pulses passing any point on the main channel is governed by the number of upstream tributary confluences. Eye-witness accounts of discharge pulses on a major tributary of the Il Eriet in the Koobi Fora Sedimentary Basin report a considerable quantity of flotsam accompanying each wave

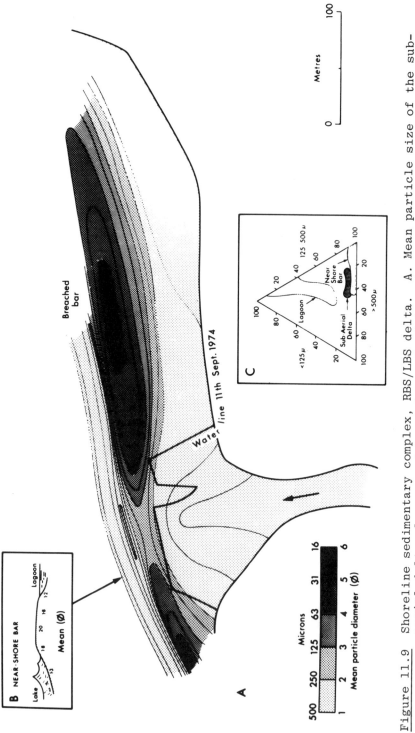

Figure 11.9 Shoreline sedimentary complex, RBS/LBS delta. A. Mean particle size of the subaerial delta, lagoon and beach bar. B. Schematic cross-section of beach bar and near-shore bar mean particle size. C. Triordinate plot of grain size for the three shoreline sedimentary environments.

Sand-bed ephemeral streams

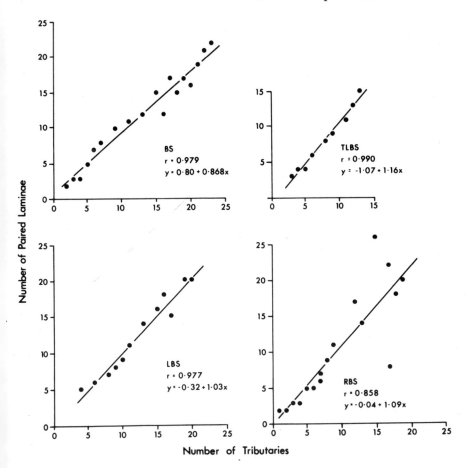

Figure 11.10 Functional relationship of the number of paired laminae and the number of upstream tributary confluences in ephemeral stream-channel fill.

(R. Watkins 1978, *personal communication*). This is taken to represent the initial sweeping of organic debris from each successive dry tributary bed.

The passage of each pulse brings with it a momentary increase in stream power and a traction carpet dominated by coarse particles. Finer particles are kept in motion while the coarser material congregates on the stream bed. As the pulse wanes, finer particles and especially the heavy, platy magnetite grains, settle, dominating the bed. Larger particles are now rejected since they find no convenient niche in the bed microtopography. In this manner each discharge pulse generates two laminae.

In order to test the general principle of drainage-net control of primary structures in ephemeral streams, the 1974 study was extended to catchments BS, LBS and TLBS (Fig 11.10). The outcome is more conclusive than the already remarkable relationship of RBS, which is included, for purposes of direct comparison, without square-root transformation - a factor which reduces slightly the correlation coefficient from that previously published (r = 0.903)

CONCLUSIONS

The elongate nature of these strike-etched Koobi Fora catchments conveniently exaggerates the role played by drainage-net configuration in shaping the flood hydrograph. The downstream changes in sediment character that result are well-defined, providing a set of diagnostic tools which are useful not only in the analysis of flood generation and the problems of sedimentation in arid-zone ephemeral streams of the present day, but also in palaeoenvironmental reconstruction.

The linear pattern of sediment size detected in catchment RBS' channel-fill contradicts the downstream fining commonly expected of fluvial deposits (Table 11.1). Although other studies report cases of coarsening bed material (Hack 1957; Brush 1961; Douglas et al 1971), only Whetton et al (1969) invoke positive change of a hydraulic variable (in their case increasing water velocity) as partial explanation of the sorting pattern. Downstream increase in stream power is not peculiar to the Koobi Fora catchments, nor to ephemeral channels in general. However, scour and fill processes permit clear identification of the deposit of a single flood event throughout the channel length, and eliminate the problem (common to most studies of perennial streams) of unravelling unlabelled but juxtaposed contributions of both high- and low-recurrence interval flood discharges.

The unexpected sorting pattern of the channel is replaced beyond the lake margin by a classical off-shore fining of particle-size. Besides the practical value of its implications, for instance in predicting the character of deltaic sedimentation in storage reservoirs on ephemeral streams (shown by Leopold et al, 1966, to be a rapid, if spasmodic, process), the subaerial and sublacustrine sorting pattern of the Koobi Fora basin provides a model for stratigraphical interpretation of those facies of the Kocbi Fora Formation thought to be the product of short ephemeral channels (Vondra & Bowen 1976). The achievement of higher stratigraphical, and therefore palaeogeographical resolution, that this provides may improve the picture we have of Early Man's environment.

As well as offering a mechanical justification of downstream coarsening in channel fill, the Koobi Fora study endows horizontal lamination for the first time with a significance in palaeoenvironmental reconstructions beyond that of merely indicating contemporary water velocities. While acknowledging that discharge pulses are not exclusively produced by the staggered superimposition of tributary contributions on main channel flow, a tally of the number of paired horizontal laminae in ephemeral stream channel-fill does add another dimension to stratigraphical interpretation, providing an indication of contemporary drainage-net configuration.

ACKNOWLEDGEMENTS

Our 1974 expedition to northern Kenya was made at the kind invitation of Richard Leakey, Director of the National Museum, and financed by the University of London Central Research Fund and Birkbeck College. The 1976 expedition forms part of a larger project carried out under Kenya Government Research Permit OP 13/OO1/6C94A and sponsored by the Natural Environment Research Council of the United Kingdom (Grant GR3/2897). We would like to thank Richard Leakey and Frank Fitch for their logistical support, and for their continued interest in our research programme.

REFERENCES

Allen, J.R.L., 1963. The classification of cross-stratified units with notes on their origin. *Sedimentology*, 2, 93-114

Allen, J.R.L., 1965. Late Quaternary Niger Delta and adjacent areas: sedimentary environments and lithofacies. *Bulletin of the American Association of Petroleum Geologists*, 49, 547-600

Axelsson, V., 1967. The Laiture Delta. *Geografiska Annaler*, 49A, 1-127

Bathurst, J.C., Thorne, C.R. & Hey, R.D., 1977. Direct measurements of secondary currents in river bends. *Nature*, 269, 504-506

Black, P.E., 1972. Hydrograph responses to geomorphic model watershed characteristics and precipitation variables. *Journal of Hydrology*, 17, 309-329

Bridge, J.S. & Jarvis, J., 1976. Flow and sedimentary processes in the meandering River South Esk, Glen Clova, Scotland. *Earth Surface Processes*, 1, 303-336

Bridge, J.S., 1977. Flow, bed topography, grain size and sedimentary structure in open channel bends: a 3-dimensional model. *Earth Surface Processes*, 2, 401-416

Bridge, J.S., 1978. Origin of horizontal lamination under turbulent boundary layers. *Sedimentary Geology*, 20, 1-16

Brush, L.M., 1961. Drainage basins, channels, and flow characteristics of selected streams in central Pennsylvania. *United States Geological Survey Professional Paper*, 282-F, 145-181

Butzer, K.W., 1971. Recent history of an Ethiopian delta. *University of Chicago, Department of Geography Research Paper*, 136

Colby, B.R., 1963. Fluvial sediments: a summary of source, transportation, deposition and measurement of sediment discharge. *Bulletin of the United States Geological Survey*, 1181-A

Crickmore, M.J., 1967. Measurement of sand tracer transport in rivers with special reference to tracer methods. *Sedimentology*, 8, 175-228

Culbertson, J.K., Scott, C.H. & Bennett, J.P., 1971. Summary of alluvial channel data from Rio Grande conveyance channel, New Mexico, 1965-1969. *United States Geological Survey Professional Paper,* 562-J

Diskin, M.H. & Lane, L.J., 1972. A basin-wide stochastic model for ephemeral stream runoff in southeastern Arizona. *Bulletin of the International Association of Scientific Hydrology,* 17, 61-76

Douglas, I., Leatherdale, J.D. & Pitty, A.F., 1960. *Unpublished* in Pitty, A.F., *Introduction to Geomorphology,* 1971 (Methuen, London)

Einstein, H.A., 1950. The bed-load function for sediment transportation in open channel flows. *Technical Bulletin of the United States Department of Agriculture,* 1026

Evans, O.F., 1939. Sorting and transportation of material in the swash and backwash. *Journal of Sedimentary Petrology,* 9, 28-31

Findlater, I.F., 1976. Tuffs and the recognition of isochronous mapping units in the East Rudolf succession. In: *Earliest man and environments in the Lake Rudolf Basin,* ed, Y. Coppens *et al,* (University of Chicago Press, Chicago), 94-104

Fitch, F.J., Watkins, R.T. & Miller, J.A., 1975. Age of a new carbonatite locality in northern Kenya, *Nature,* 254, 581-583

Fox, W.T., Ladd, J.W. & Martin, M.K., 1966. A profile of the four moment measures perpendicular to a shoreline, South Haven, Michigan. *Journal of Sedimentary Petrology,* 36, 1126-1130

Francis, J.R.D., 1973. Experiments on the motion of solitary grains along the bed of a water-stream. *Proceedings of the Royal Society, Series A,* 332, 443-471

Francis, J.R.D. & Asfari, A.F., 1971. Velocity distributions in wide, curved open-channel flows. *Journal of Hydraulic Research,* 9, 73-90

Frostick, L.E. & Reid, I., 1977. The origin of horizontal laminae in ephemeral stream channel-fill. *Sedimentology,* 24, 1-9

Gregory, K.J. & Walling, D.E., 1973. *Drainage basin form and process* (Arnold, London)

Hack, J.T., 1957. Studies of longitudinal stream profiles in Virginia and Maruland. *United States Geological Survey Professional Paper,* 294-B

Harms, J.C. & Fahnestock, R.K., 1965. Stratification, bedforms and flow phenomena (with an example from the Rio Grande), *Special Publication of the Society of Economic Paleontologists & Mineralogists,* 12, 192-219

Hooke, R. LeB., 1975. Distribution of sediment transport and shear stress in a meander bed. *Journal of Geology,* 83, 543-565

Jopling, A.V., 1964. Laboratory study of sorting processes related to flow separation. *Journal of Geophysical Research*, 69, 3403-3418

Kennedy, V.C. & Kouba, D.L., 1970. Fluorescent sand as a tracer of fluvial sediment. *United States Geological Survey Professional Paper*, 562-E

Kolmer, J.R., 1973. A wave-tank analysis of the beach foreshore grain-size distribution. *Journal of Sedimentary Petrology*, 43, 200-204

Kuenen, Ph.H., 1966. Experimental turbidite lamination in a circular flume. *Journal of Geology*, 74, 523-545

Laronne, J.B. & Carson, M.A., 1976. Interrelationships between bed morphology and bed material transport for a small gravel-bed channel. *Sedimentology*, 23, 67-85

Leakey, R.E.F., 1970. New hominid remains and early artefacts from northern Kenya. *Nature*, 226, 223-224

Leakey, R.E.F., 1976. An overview of the Hominidae from East Rudolf, Kenya. In: *Earliest man and environments in the Lake Rudolf Basin*, ed. Y. Coppens et al, (University of Chicago Press, Chicago), 476-483

Leopold, L.B. & Miller, J.P., 1956. Ephemeral streams - hydraulic factors and their relation to the drainage net. *United States Geological Survey Professional Paper*, 282-A

Leopold, L.B., Emmett, W.W. & Myrick, R.M., 1966. Channel and hillslope processes in a semi-arid area, New Mexico. *United States Geological Survey Professional Paper*, 352-G, 193-253

Martinec, I.J., 1967. The effect of morphological processes in alluvial channels on flow conditions. *Publication of the International Association of Scientific Hydrology*, 75, 243-248

Mathews, W.H. & Shepard, F.P., 1962. Sedimentation of Fraser River Delta, British Columbia. *Bulletin of the American Association of Petroleum Geologists*, 46, 1416-1443

McBride, E.F., Shepherd, R.G. & Crawley, R.A., 1975. Origin of parallel, near horizontal laminae by migration of bedforms in a small flume. *Journal of Sedimentary Petrology*, 45, 132-139

McKee, E.D., Crosby, E.J. & Berryhill, H.L., 1967. Flood deposits, Bijou Creek, Colorado, June 1965. *Journal of Sedimentary Petrology*, 37, 829-851

Meland, N. & Norrman, J.O., 1966. Transport velocities of single particles in bedload motion. *Geografiska Annaler*, 48A, 165-182

Mississippi River Commission, 1935. Studies of river bed materials and their movement with special reference to the lower Mississippi River. *Paper of the United States Waterways Experimental Station*, 17

Moss, A.J., 1963. The physical nature of common sandy and pebbly deposits. Part II. *American Journal of Science*, 261, 297-343

Müller, G., 1966. The new Rhine delta in Lake Constance. In: *Deltas*, ed. M. L. Shirley & J. A. Ragsdale (Houston Geological Society), 107-124

Naidu, A.S., 1966. Lithological and chemical facies changes in the recent deltaic sediments of the Godavari River, India. In: *Deltas*, ed. M. L. Shirley & J. A. Ragsdale, (Houston Geological Society), 125-157

Normark, W.R. & Dickson, F.H., 1976. Sublacrustrine fan morphology in Lake Superior. *Bulletin of the American Association of Petroleum Geologists*, 60, 1021-1036

Pettijohn, F.J., 1957. *Sedimentary rocks*. (Harper, New York)

Picard, M.D., & High, L.R., 1973. Sedimentary structures of ephemeral streams. *Developments in Sedimentology*, (Elsevier, Amsterdam), 17

Plumley, W.J., 1948. Black Hills terrace gravels: a study in sediment transport. *Journal of Geology*, 56, 526-577

Pollack, J.M., 1961. Significance of compositional and textural properties of South Canadian River channel sands, New Mexico, Texas and Oklahoma. *Journal of Sedimentary Petrology*, 31, 15-37

Potter, P.E., 1955. The petrology and origin of the Lafayette Gravel. Part I - Mineralogy and Petrology. *Journal of Geology*, 63, 1-38

Renard, K.G. & Keppel, R.V., 1966. Hydrographs of ephemeral streams in the Southwest. *Journal of Hydraulics Division, American Society of Civil Engineers*, 92(HY), 33-52

Renard, K.G., 1970. The hydrology of semi-arid rangeland watersheds. *Publication of the United States Department of Agriculture, Agricultural Research Series*, 41-162

Roberts, M.C. & Klingeman, P.C., 1970. The influence of landform and precipitation parameters on flood hydrographs. *Journal of Hydrology*, 11, 393-411

Sanders, J.E., 1965. Primary sedimentary structures formed by turbidity currents and related resedimentation mechanisms. *Special Publication of the Society of Economic Palaeontologists and Mineralogists*, 12, 192-219

Sharon, D., 1972. The spottiness of rainfall in a desert area. *Journal of Hydrology*, 17, 161-175

Shepard, F.P., 1956. Marginal sediments of Mississippi delta. *Bulletin of the American Association of Petroleum Geologists*, 40, 2537-2623

Smith, N.D., 1970. The braided stream depositional environment: comparison of the River Platte with some Silurian clastic rocks, north-central Appalachians. *Bulletin of the Geological Society of America*, 81, 2993-3014

Smith, N.D., 1971. Pseudo-planar stratification produced by very low amplitude sand waves. *Journal of Sedimentary Petrology*, 41, 69-73

Smith, N.D., 1974. Sedimentology and bar formation in the Upper Kicking Horse River, a braided outwash stream. *Journal of Geology*, 82, 205-223

Thornes, J.B., 1976. Semi-arid erosional systems: case studies from Spain. *London School of Economics Geography Paper*, 7

Vondra, C.F. & Bowen, B.E., 1976. Plio-Pleistocene deposits and environments, East Rudolf, Kenya. In: *Earliest man and environments in the Lake Rudolf Basin,* ed. Y. Coppens *et al,* (University of Chicago Press, Chicago), 79-93

Ward, R.C., 1975. *Principles of Hydrology,* (McGraw-Hill, London), 2nd edition

Whetten, J.T., Kelley, J.C. & Hanson, L.G., 1969. Characteristics of Columbia River sediment and sediment transport. *Journal of Sedimentary Petrology*, 39, 1149-1166

Wooding, R.A., 1966. A hydraulic model for the catchment-stream problem. III - Comparison with runoff observations. *Journal of Hydrology*, 4, 21-37

Wright, L.D. & Coleman, J.M., 1974. Mississippi River mouth processes: effluent dynamics and morphological development. *Journal of Geology*, 82, 751-778

Yen, B.C. & Chow, V.T., 1969. A laboratory study of surface runoff due to moving rainstorms. *Water Resources Research*, 5, 989-1006

CHAPTER 12 THE ANALYSIS OF PEBBLES ON THE BED OF THE UPPER WHARFE, YORKSHIRE

P. A. Whittel[*]

INTRODUCTION

Point bar deposits and river bed gravels have been extensively studied in the past 60 years, following early sedimentological work by C. K. Wentworth (1919). His studies of the pebble fraction of deposits of marine (Wentworth 1922a) as well as fluvial environments (Wentworth 1922b) directed most of the main lines of subsequent investigations in North America. Later, European workers J. Tricart and A. Cailleux (1959) comprehensively considered the evidence for discerning effects of distinctive climatic environments on pebble morphometry. Throughout there has been a very active interest in the definition of indices by which pebbles might be most effectively described and from which the significance of varying geomorphological environments might be specified. For instance, C. K. Wentworth (1922 a, b) defined ratios of roundness and of flatness. From the plotting of these two indices he distinguished three groups of pebbles which appeared to indicate a glacial, fluvial or a sand-blasted origin. He also plotted the progressive change in pebble shape according to increasing length of time during which test pebbles were subject to an artificial, tumble mill process. Concurrently he analysed downstream pebble shape development on Russell Fork on the Big Sandy River, Virginia. Rounding was rapid over the first 5 kms of the stream channel, but with only very gradual increases over a long distance further downstream.

These studies were extended mainly by American workers in semi-arid or arid fluviatile environments in the USA, including studies of the flood deposits of the Arroyo Seco, Los Angeles County, and the San Gabriel Canyon, California (Krumbein 1940, 1941a, 1942). Tributary streams were found to be the major control on percentage type of pebble, although this influx did not affect the increase of pebble roundness which was attributed to abrasion during transit along the main valley. Large cobbles in Brandywine Creek, Maryland, were analyzed to ascertain whether they were fluviatile or glacial in origin (Krumbein & Lieblien 1956) and the effects of geology on channel characteristics and its material in New Mexico were studied (Miller 1958). Elsewhere extensive work was undertaken in Europe and also in the humid tropics (Tricart & Schaeffer 1950; Tricart & Cailleux 1959; Tricart & Vogt 1967). In northern Italy

---
[*] Birkenhead School

R. Dal Cin (1968) has studied the downstream characteristics of pebbles in the rivers Piave and Avisio, particularly their percentage composition and roundness variations. A. Cailleux and J. Tricart's roundness and flatness ratios have been used in an attempt to establish modes of transportation in streams during passage over post-Cretaceous formations in north-east Brazil (Mabesoone 1966). In Australia the downstream rounding of sediments in the Hacking River, New South Wales, was unaffected by tributary confluences because of their relatively small size and petrographic uniformity of the river catchment (Ouma 1967). However, tributary entrance usually complicates the picture by introducing a variety of new materials (Leopold et al 1964). Indeed, in many valleys such as the Wabash valley in Indiana, the possible mutual interactions affecting the selection, abrasion and sorting of bedload may include contributions from tributary streams, glacial materials and bedrock too complicated for discernible trends to emerge (McCammon 1961). Clearly if downstream changes in the proportions of a given rock fragment are to be identified, and the change attributed to either a mechanical destruction or simply to a dilution of these fragments, a study area of reasonably uncomplicated geology is essential.

The purpose of the present investigation is to establish a case study for the British Isles which is comparable with preceding work elsewhere in which noteworthy trends have been established, but on a scale typical of the size of British rivers. Upper Wharfedale is well-suited for such a purpose, not simply because it is well-known, but mainly because the geological succession and structural disposition of the area are constant or constantly changing factors, and the major contrast, introduced by the Mid-Craven fault, is clear-cut and decisive.

GEOLOGY AND RELIEF OF UPPER WHARFEDALE

As a result of geological variations, upper Wharfedale can itself be sub-divided. An upper section of the valley which drains rocks which form the Askrigg Massif (Hudson 1938) and the lower part, downstream from Grassington, the Bowland Trough (Fig 12.1). On the Massif, the Carboniferous rocks of the Millstone Grit, Yoredale Series, and the Great Scar Limestone divisions are exposed. As shown in the geological cross-sections 1-6 of Fig 12.1, the headwater area is dominated by shales, sandstones and limestones. The Millstone Grit forms a capping to the valley sides, the Yoredale Series being predominant in the headwaters but giving way to Great Scar Limestone down-valley as it thins out with increasing distance from the type-area to the north in Wensleydale (Phillips 1836). Downstream the valley is cut progressively deeper into the Great Scar Limestone, in part an expression of the north-easterly regional dip of about 1:100. As shown in section 7 (Fig 12.1) the Great Scar Limestone (Rayner 1953) dominates the Wharfe catchment at the southern edge of the massif. Section 8 near Grassington marks the boundary of

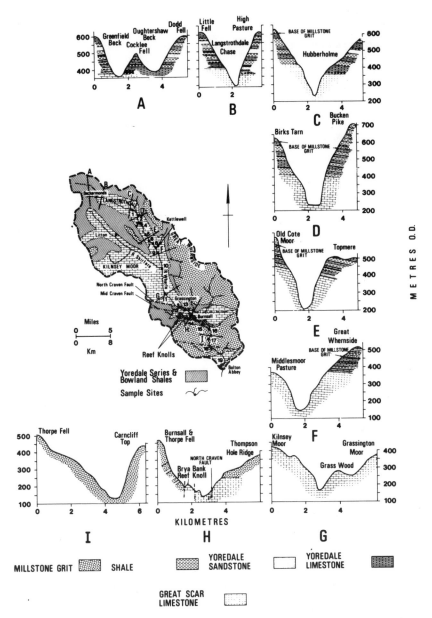

Figure 12.1 Geological characteristics of Upper Wharfedale and position of pebble-sampling localities (The geological information is based on unpublished 6-inch maps in the Leeds Office of the Institute of Geological Sciences, a division of the Natural Environment Research Council. Permission to publish this geological information is gratefully acknowledged.)

the Massif at the North and Mid-Craven Fault systems (Wager 1931). Downstream, the Millstone Grit forms the Wharfe valley (Figure 12.1, section 9), providing a major downstream change in lithology. Therefore the effect of the varying significance of the factors influencing river pebble characteristics will be in parts both subtle and abrupt.

PEBBLE BEDLOAD SAMPLING

A commonly followed sampling procedure suggests that a number of metre-square grids be randomly selected at several locations on a point-bar deposit (Wolman 1954). This procedure ensures that the characteristics of *one* point bar are comprehensively established but, without modification, the procedure is not suited to an investigation in which it is the general geographical trend of several point-bars downstream which is the focus of interest. Another factor is that American workers commonly have studied ephemeral streams during the dry season with the bedload fully exposed on the dry channel floor. The perennial flow of the Wharfe necessitated that the main priority for present purposes was the need to standardise the height of water level. Thus, sampling took place over a period of a few days when river stage was constantly re-checked to ensure that no change in flow occurred. In all 19 point bars were visited in upper Wharfedale (Fig 12.2) spaced at approximately equal distances (Fig 12.1). Sample points were established at the midpoint of each bar along the line of the water mark. From the surface pebbles 100 were collected at each bar from an area $1m^2$ on the landward side of the water mark. A non-specialist assistant picked the pebbles as quickly as possible so that pre-conception or forethought could add little bias to the selection.

DOWNSTREAM CHANGES IN RELATIVE PROPORTIONS OF ARENACEOUS AND LIMESTONE PEBBLES

From Hubberholme to Kettlewell there is an overall downstream fall in the percentage of the arenaceous pebble fraction from 44 per cent to 16 per cent. This reflects the strengthening of the limestone sequences in the Yoredale cyclothems away from the Yoredalian type area in Wensleydale. In addition, the extent of Millstone Grit, as a caprock on the plateau tops in upper Wharfedale decreases down this portion of the valley. A second feature of the data is that its scatter about the general trend of a 28 per cent decrease in arenaceous pebbles over 8 kms, narrows downvalley. This tendency for the general trend to become more clearly established is possibly an expression of the progressive decline in the importance of tributary streams on the main channel as it increases in dimensions and thus in relative importance. A similar characteristic was noted in the flood gravels of San Gabriel Canyon, California (Krumbein 1940).

Downstream from Kettlewell, the Yoredale sequence continues to thin, and due to the prevailing regional dip of about 5^0 to the north and north-east there is an

Figure 12.2 Point bar on the River Wharfe, 1 km downstream from Burnsall. This is sample site 14: river flow is from right to left.

Figure 12.3 Gravel at sample site 14, showing the position of the sampling grid.

Pebbles from the Upper Wharfe

increasing preponderance of Great Scar Limestone brought up to form the valley sides and to flank tributary valleys. Areas like Kilnsey Moor and Grassington Moor to the east thus contribute chiefly limestone materials to the fluvial system. The large tract of Millstone Grit on Grassington Moor is part of a watershed of the southward draining Hebden Beck and Barden Beck which does not drain to the south-west to join the Wharfe until downstream from Grassington (Fig 12.1). Thus in the main channel between Kettlewell and Grassington the proportion of arenaceous rock pebbles is halved from 16 per cent to 7 per cent.

In a third section in the present study area the Wharfe crosses the mid-Craven fault, and Millstone Grit abruptly becomes the bedrock and that of the tributary streams draining the flanks of the main valley. From this point, Millstone Grit is the only rock type added to the channel and the percentage of arenaceous pebbles increases from 7 per cent to 46 per cent between Grassington and Bolton Abbey. The 8 points from the point bar samples exhibit a linear trend (Fig 12.4) and might suggest some equilibrium in the fluvial processes in this section of the valley with its tributary contributions. The fluvial process of abrasion wears the limestone down, whilst simultaneously the Millstone Grit is added from the tributary streams below Burnsall. The downstream decrease in limestone of 35 per cent in 9.7 kms (3.6 per cent/km) compares closely with the reversed relationship in the upstream Hubberholme-Kettlewell reach where the arenaceous fraction fell 28 per cent in 8 kms (3.5 per cent/km).

These trends provide an interesting basis for comparison with findings in other areas and from different lithologies. For instance, in the Black Hills of Dakota a decrease of 23.6 per cent in limestone and sandstone pebbles was observed over a distance of 48.2 km (Plumley 1948). In the Lower Colorado a 70 per cent limestone concentration in the bedload diminished to less than 1 per cent within a distance of less than 250 km (Bradley 1970). In the Merced River, Sierra Nevada, the granite fraction in pebbles decreased from 100 per cent to 3 per cent within 43 km (Pittman & Ovenshine 1968) although, in the Lower Colorado, a 24 per cent granite proportion of bedload deposits took 250 km to reduce to a 2 per cent proportion, and 400 km before it disappeared entirely (Sneed & Folk 1958). Indeed, in the cases of particularly durable rocks, no loss of fragments may be observed. For example, in the Elk Creek in the Black Hills, no loss of felsitic rock fragments was detected over a distance of 160 km, although there was a definite increase in roundness (Cameron & Blatt 1971).

PEBBLE DURABILITY

1. Methods

The changes of the relative proportion of rock types in the bedload of the River Wharfe are consistent with a clearly defined geological control. On the other hand, certain striking consistencies in the results might be

attributed to the efficiency of the fluvial process. In
consequence it was decided to investigate the durability of
the pebble load as differential rates of abrasion may also
be involved as well as a simple winnowing or enrichment,
according to the bedrock in the tributary areas.

It is well-established from the tumble-mill and similar
experiments that tests on pebble response to fluvial
transportation are feasible only if simulated in artificial
conditions (Kuenen 1956). Therefore, standardised
empirical tests for estimating wearing properties of
aggregates (Ministry of Transport 1952) which have proved
satisfactory in engineering practice were considered for
the present scientific enquiry and the Modified Aggregate
Impact Test (Hosking & Tubey 1969) was adopted. This test
involves the use of a 30 lb hammer falling from a height
of 15" on to the test pebble. Pebble sizes were
standardised and it was found that the most suitable
sample was a pebble with a long axis dimension of 100 mm.
The mean size of pebbles at any one sample site approxi-
mated to 58 mm in median dimension, with a between size
mean variation of $\pm$ 4 mm. Fragmentation of the pebble
sizes chosen yielded a sufficient percentage of 'fines'
which are defined as that weight of crushed pebble fragments to pass through a No. 8 sieve (2.14 mm mesh). Since
the absolute quantity of fines is clearly an artificial
product of the measuring test, the only significant
conclusion to be drawn is to note their consistency with
values quoted in the standard manual (Lovegrove 1929).
Otherwise any scientific interpretation depends on the
significance of the relative proportions of fines from one
geographical locality to another.

2. Discussion of pebble durability data

In upper Wharfedale, between Hubberholme and Kettlewell,
the limestone pebbles show a consistency in durability with
only very minor local fluctuations. They are essentially
very durable pebbles, yielding consistently fewer than
10 per cent fines. However, downstream from Kettlewell
there is a gradual progressive increase in percentage
fines such that their value at Bolton Abbey is twice that
at Kettlewell. The inference that limestone pebbles by
the time they had been carried as far as Bolton Abbey
were twice as susceptible to impact abrasion is consistent
with the known mechanical properties of crystalline
limestone, in which the three perfect cleavage planes of
calcite introduce a certain mechanical weakness into the
rock which is then susceptible to fragmentation under
impacts beyond a certain force.

In contrast to the limestone the arenaceous materials
appear to be less durable than the limestones and to show
more marked downstream changes. Initially the pebble contribution in the headwater areas has virtually the same
resistance to impact as limestones, but susceptibility
doubles progressively in the Hubberholme-Kettlewell reach.
The subsequent increase over the Kettlewell-Grassington
reach is much less marked, but parallels almost exactly the
slight but progressive increase in the susceptibility to

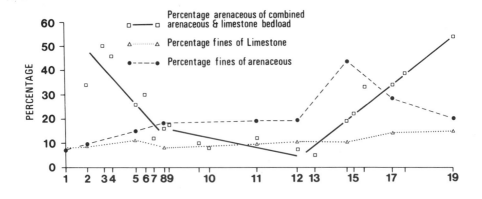

Figure 12.4 Downstream changes in pebble composition and durability in the Upper Wharfe. The greater the percentage of fines, the less durable the bedload as described in the text. The length of the reach sampled is 24 km and the position of sampling stations is plotted to scale.

impact demonstrated by the limestone pebbles. The striking increase in susceptibility to impact of the arenaceous materials at Burnsall is no more noteworthy than the subsequent decline with fines of 43.7 per cent being reduced to 20.3 per cent at Bolton Abbey. There is no known change in Millstone Grit lithology in the country rock in this direction which could account for an increase in relative durability. On the other hand, the durability does not approach that observed in the headwater area at Hubberholme. In fact by Bolton Abbey the arenaceous material is no greatly more susceptible to impact than that which has been gradually attained by the limestone portion. Clearly some channel process is perhaps the explanation, but this could not be preponderantly a mechanical effect otherwise the rate of increase in the susceptibility of limestone pebbles to impact would also decline instead of maintaining its steady increase over this lowest reach in the study area.

An important feature of Upper Carboniferous arenaceous rocks in relation to their durability is their porosity. The Yorkshire sandstones of the Millstone Grit series are often very siliceous, but they are typically porous and their grains are not firmly cemented together (Lovegrove 1929). One possibly significant environmental factor is the chemistry of the river water which would saturate both the porous sandstone and the permeable limestone and affects the type of mosses which colonise gravel temporarily immobilised above annual flood levels. In the upper Wharfe, alkalinity increases as far as Grassington where lime-rich groundwaters are forced to the surface near the Craven fault. Downstream alkalinity declines from pH = 7.65 at Burnsall to pH = 7.17 at Barden Bridge and finally pH = 7.09 at Bolton Abbey. However, it would clearly be premature to speculate about the possible significance of these or other factors before a geographically more intensive sampling established more precisely the nature of downstream changes along what Fig 12.4 reveals as critical reaches.

CONCLUSIONS

Despite the difficulties of demonstrating relationships in pebble lithologies amid the complexities of real river channels, it is readily assumed that pebbles are modified to varying degrees and at different rates by downstream transport. In the case of upper Wharfedale it seems that the spatial variations in sandstone and limestone gravel characteristics of the stream bedload are influenced chiefly by the geological factors of lithology and structure, and only secondarily by fluvial processes. This conclusion agrees with that based on the study of the Black Hills terrace gravels where "... initial lithological frequency distribution of a gravel is directly related to the source area which furnishes detritus to a stream" (Plumley 1948). The changing catchment geology is apparently the most important factor, with the different responses of the lithological pebble types to the hydro-physical processes being superimposed to a much lesser degree on the pebble composition. Such changes, apart from their significance in sedimentology and hydraulics, are of interest to the geomorphologist not simply because river bedload is an integral part of river channel processes, but also because its depositional forms make up significant portions of the stream channel and banks.

REFERENCES

Bradley, W.C., 1970. Effects of weathering on abrasion of granitic gravel, Colorado River, Texas. *Bulletin of the Geological Society of America,* 81, 61-80

Dal Cin, R., 1968. Climatic significance of roundness and percentage of quartz in conglomerates. *Journal of Sedimentary Petrology,* 38, 1094-1099

Hosking, J.R. & Tubey, L.W., 1969. Research on low grade and unsound aggregates. *Ministry of Transport, Road Research Laboratory Report,* LR293, 1-35

Hudson, R.G.S., 1938. The Carboniferous rocks. In: The geology of the country around Harrogate. *Proceedings of the Geologists' Association,* 49, 306-330

Krumbein, W.C., 1940. Flood gravel of San Gabriel Canyon, California. *Bulletin of the Geological Society of America,* 51, 639-676

Krumbein, W.C. & Lieblien, J., 1956. Geologic application of extreme-value methods to interpretation of cobbles and boulders in gravel deposits. *Transactions of the American Geophysical Union,* 37, 313-319

Kuenen, P.H., 1956. Experimental abrasion of pebbles: 2. Rolling by current. *Journal of Geology,* 64, 336-368

Leopold, L.B., Wolman, M.G. & Miller, J.P., 1964. *Fluvial processes in geomorphology.* (Freeman, San Francisco and London)

Lovegrove, E.J., 1929. Attrition tests of British roadstone. *Memoir of the Geological Survey of the United Kingdom.*

Mabesoone, J.M., 1966. Relief of north-eastern Brazil and its correlated sediments. *Zeitschrift für Geomorphologie,* 10, 419-453

McCammon, R.B., 1961. Variation in pebble composition of Wisconsin outwash sediments in the Wabash Valley, Indiana. *Journal of Sedimentary Petrology,* 31, 73-79

Miller, J.P., 1958. High mountain streams: effects of geology on channel characteristics and materials. *Memoir of the New Mexico State Bureau Mines and Mineral Resources,* 4

Ministry of Transport, 1952. *Soil mechanics for road engineers.* Government Publications, Her Majesty's Stationery Office

Ouma, J.P.B.M., 1967. Fluviatile morphogenesis of roundness: the Hacking River, New South Wales, Australia. *Publication of the International Association of Scientific Hydrology,* 75, 319-343

Phillips, J., 1836. *Illustrations of the geology of Yorkshire. Part II. Mountain Limestone district.* (Murrays, London)

Pittman, E.D. & Ovenshine, A.T., 1968. Pebble morphology in the Merced River (California). *Sedimentary Petrology,* 2, 125-140

Plumley, W.J., 1948. Black Hills terrace gravels: study in sedimentary transport. *Journal of Geology,* 56, 526-578

Rayner, D.H., 1953. The Lower Carboniferous rocks in the north of England (a review). *Proceedings of the Yorkshire Geological Society,* 28, 231-315

Sneed, E.D. & Folk, R.L., 1958. Pebbles in the Lower Colorado valley, Texas. A study in particle morphogenesis. *Journal of Geology,* 66, 114-150

Tricart, J. & Cailleux, A., 1959. *Initiation a l'etude des sables et des galets.* (Centre de Documentation Universitaire, Paris)

Tricart, J. & Schaeffer, R., 1950. L'indice d'emousse des galets, moyen d'etude des sytemes d'erosion. *Revue de Geomorphologie dynamique,* 1, 151-179

Tricart, J. & Vogt, H., 1967. Quelques aspects du transport des alluvions grossières et du faconnement des lits fluviaux. *Geografiska Annaler,* 49A, 351-366

Wager, L.R., 1931. Jointing in the Great Scar Limestone of Craven and its relation to the tectonics of the area. *Quarterly Journal of the Geological Society,* 87, 392-424

Wentworth, C.K., 1919. A laboratory and field study of cobble abrasion. *Journal of Geology,* 27, 507-521

Wentworth, C.K., 1922a. The shape of beach pebbles. *United States Geological Survey Professional Paper,* 131-C, 74-83

Wentworth, C.K., 1922b. A field study of shapes of river pebbles. *Bulletin of United States Geological Survey,* 730, 103-114

Wolman, M.G., 1954. A method of sampling coarse river-bed material. *Transactions of the American Geophysical Union,* 35, 951-956

CHAPTER 13 THE MORPHOLOGICAL RELATIONSHIPS OF BENDS IN CONFINED STREAM CHANNELS IN UPLAND BRITAIN

J. A. Milne[*]

INTRODUCTION

At one extreme a landform may be a crag of unweathered rock. At the other it is a loose pile of sand which shifts with each tide or flood. In stream courses the immediate surroundings of the channel change gradually downstream with stream flow increasingly, but reversibly, dominating environmental resistance. The tendency downstream is for bed-material size and channel gradient to decrease, unconsolidated materials are more susceptible to fluvial re-working, and the lateral, valley-side constraints are reduced. The downstream increase in discharge compensates for the decrease in channel gradient and transportation becomes increasingly that of suspended sediment.

In headstream areas the land surface is largely one of bedrock beneath a thin regolith of drift cover, and the plan of first and second order gullies is essentially straight. Downstream, unconsolidated valley-floor sediments and significant decrease in channel gradient allow some increase in the sinuosity of larger channels, although valley walls of rock or unconsolidated valley-fill commonly remain as significant restraints. Along reaches confined by valley walls, conditions may locally favour the development of regular meanders and associated cross-sectional forms. Generally, however, the likelihood of bedrock outcrops is too common and the valley-floor materials are too coarse and variable spatially to allow fluvial processes to overcome the resistance offered by the environment for any distance. Thus, most cross-sections and meander plans in confined stream channels in upland Britain are irregular in shape. Beyond the uplands trunk streams, generally greater than 6th order, gradually become free to move laterally without any bedrock constraints. The change from an upland to a lowland environment is usually coincident with a change in channel gradient sufficient to reduce gravel transport. The contrast between the gravel-bed channels of upland catchments and the sand-bed of lowland areas can be quite dramatic as seen at the emergence of rivers such as the Dove, Rye and Derwent from the North York Moors on to the Vale of Pickering.

Generalizations describing the inter-relationships between cross-section and meander size, and with flow magnitude and environmental factors have been widely

---

[*] Department of Geography, University of Newcastle-on-Tyne

Figure 13.1 The Kingledoors Valley, looking downstream, showing the development of the floodplain and the confining effect of the valleyside bluffs

established for the large, unconfined sand-bed channels in the United States (Leopold & Wolman 1957, 1960; Carlston 1965; Schumm 1967). For instance, the greater the discharge which such channels accommodate, the wider they are. This relationship indirectly explains the positive correlation between flow and bend size. Generally meander wavelengths measured across a floodplain tend to be 10-14 times greater than bankfull width and pool-riffle spacing is about half this. Lack of correlation between lateral bend development, expressed by meander amplitude, and either width or discharge, has highlighted the geographically variable extent to which the environment permits meandering to develop. Therefore the purpose of the present enquiry is to examine the degree of applicability of these generalisations to upland environments where the intervention of environmental influence is clearly evident.

THE REACHES SURVEYED

Plan-form and cross-section measurements (Table 13.1) have been made on several dozen stream bends in 11 reaches confined in uplands (Figs 13.2 and 13.3). In all some 600 cross-sections were measured. The large number of measurements established average values decisively. The streams are drawn from four geographical groups (Fig 13.3) which broader, reconnaisance field work suggested would represent the range of characteristics found in upland channels in Southern Scotland and Northern England.

1. Criteria for selecting the reaches surveyed

The geographical location and limits of the study areas were carefully chosen so that the influences of significant

Table 13.1 General characteristics of the reaches surveyed and their drainage basins. The total number of cross-sections surveyed is included together with the total number of surveyed points. The 'mean annual flood' is derived indirectly using drainage basin characteristics and climatic variables in a multiple regression equation (Natural Environment Research Council, 1975).

Reach No.	River name	Hydrometric area	Basin area km²	Basin order	Cross sections total	Survey points total	Mean annual flood (m³s)	Geology
R1	Kingledoors Burn	21 Tweed	12.49	3	60	427	9.81	Silurian and Ordovician greywackes and shales; glacial and fluvioglacial drift
R2	Stanhope Burn	21 Tweed	15.88	4	77	497	12.98	Recent alluvium
R3	Black Water (Butterholes Bridge Gap)	80 Dee Galloway	20.16	4	99	759	23.36	Silurian greywackes; glacial drift, fluvioglacial bluffs
R4	Black Water (Catherines Pool Reach)	80 Dee Galloway	18.37	3	85	612	15.03	(Recent alluvium in R3)
R5	Polmaddy Burn	80 Dee Galloway	25.53	4	37	330	37.26	Silurian greywackes etc; some granite
R6	Carsphairn	80 Dee Galloway	17.06	4	31	250	17.75	
R7	Snaizeholme Beck	27 Ouse Yorkshire	11.49	4	43	314	22.90	Yoredale sandstones, shales, limestones; drift
R8	River Ure	27 Ouse Yorkshire	17.47	4	57	408	22.58	
R9	River Rede	22 Coquet	20.22	5	46	321	29.60	U. Carboniferous sandstones, shales and limestones; glacial drift
R10	Bowmont Water	21 Tweed	70.76	5	19	169	42.88	Devonian andesite lavas, some sandstone; Silurian greywackes and shales; glacial drift
R11	Halter Burn	21 Tweed	11.49	4	21	174	9.98	

Bends in confined streams

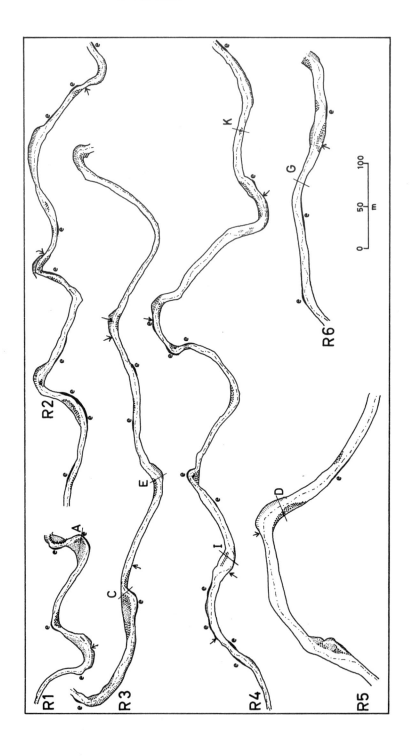

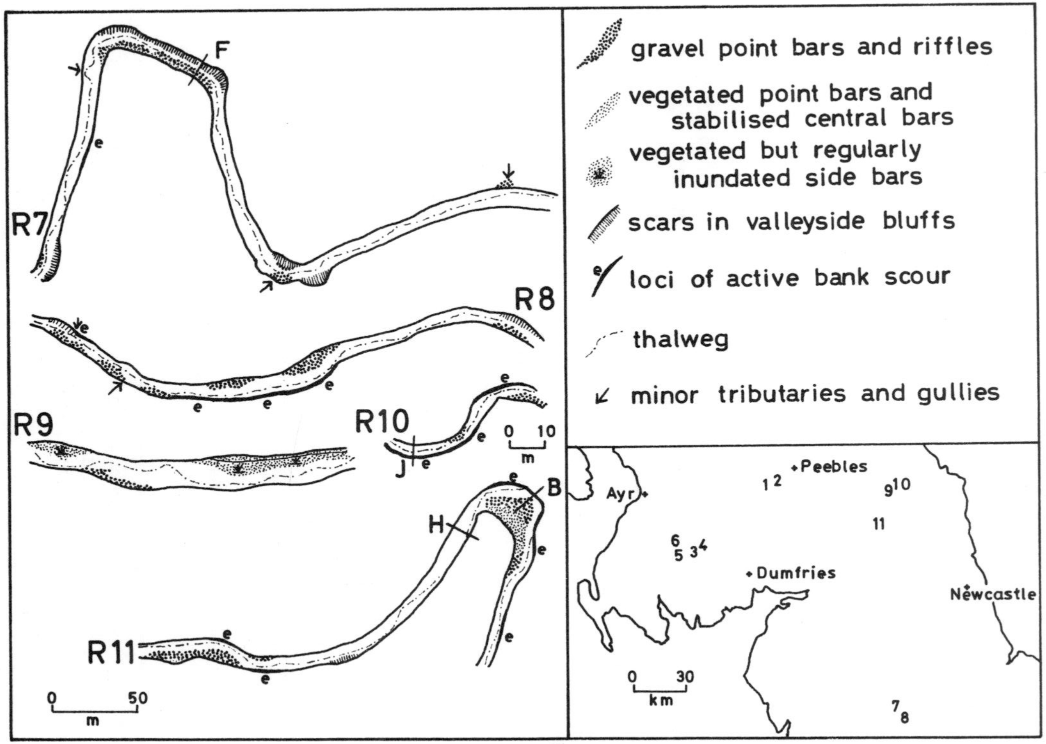

Figure 13.2 (page 228 above)
Significant features of the reaches surveyed, R1-R6. Flow is from left to right. The reaches, followed by the Grid Reference for their upstream ends are: R1 Kingledoors Burn (NT 093272); R2 Stanhope Burn (NT 131288); R3 Black Water - Catherines Pool Reach (NX 646887); R4 Black Water - Butterholes Bridge Reach (NX 640887); R5 Polmaddy Burn (NX 576887); R6 Carsphairn Lane (NX 538957). All reaches are drawn at the same scale.

Figure 13.3 (left)
Significant features of the reaches surveyed, R7-R11, and location of all reaches. R7-R11 are: R7 River Ure (SD 804926); R8 Snaizeholme Beck (SD 833879); R9 Bowmont Water (NT 837297); R10 Halter Burn (NT 838298); R11 River Rede (NT 719045). All reaches are drawn to the same scale, except R10.

factors could be highlighted. First, local geology may influence the broad patterns of valley shapes, and the size, shape and durability of much bed material. Silurian and Ordovician sedimentary rocks, for instance, may yield markedly flat pebbles. However, the presence of large amounts of fluvioglacial and glacial drift in the fluvial deposits in the channels and along channel banks minimizes the effect of the range of underlying rock types on the immediate contact of these channels with their environment. In addition, the avoidance of reaches interrupted by bedrock or artificial structures explains why the reaches measured vary in length, with certain reaches, such as those on the Kingledoors and Stanhope Burns or on the Black Water, exemplifying conditions of long, uninterrupted lengths. Secondly, certain reaches were chosen where morphological characteristics of particular interest are conspicuously displayed. Carsphairn Lane and Snaizeholme Beck were deliberately chosen as, respectively, low and high width-depth ratio channels in their cross-sections, and with a low sinuosity. Similarly, the reach along the River Ure and that along Polmaddy Burn represents the same cross-sectional contrasts, but in an environment influenced by large valley bends. To ensure some homogeneity in the size range and environment of the meanders, all surveyed reaches are located on wide, flat valley floors, confined within deep valleys and at some distance below the point where headwater gullies amalgamate (Fig 13.1). At these locations flows are sufficient to produce channels which express the regularity inherent in the fluvial process, yet still within an environment which remains, in part, resistant to erosion and scour. Halter Burn and Bowmont Water represent the two extremes of scale encompassed in the size range of meanders that it was feasible to survey in the field. Channel gradients varied little, with six surveyed reaches falling in the 0.007-0.010 range, with Stanhope Burn being the steepest (0.024).

2. <u>Local channel characteristics</u>

Several different environmental characteristics of channel perimeters can be distinguished. First, non-cohesive materials in channel banks, point bars and along the beds are coarse and poorly sorted. The average intermediate diameter of 600 bed particles along the Kingledoors Burn was 40.1mm (-5.32ϕ). In Bowmont Water, with the smallest, best-sorted bed material, the intermediate particle diameter was 24.9 mm (-4.64ϕ). The coarseness of the bed is paralleled by a lack of sand sizes on the channel bars, particularly where the bed material is derived from fine-grained Silurian mudstones and slates. There is a greater proportion of sand in the bars of streams flowing over Carboniferous strata. The reaches also have well-developed pool-riffle sequences similar to those described in the literature (Keller 1971, 1972). Secondly, important environmental changes occur where the channels impinge on the valley side. If valley-fill materials are present at such points, large scars may be cut, up to 10 m in height (Fig 13.1). Such valley deposits are poorly sorted, with a wide range of individual particles up to boulder size incorporated in a matrix of fines. In central Southern Scotland their Quaternary history has been studied in detail (Price 1963). Thirdly, the present channel may encounter distinctly cohesive deposits, the infillings of relict channels which criss-cross the floodplain. The cohesiveness of such deposits is attributable to their high percentage of silt and clay. Finally, in straight reaches the degree of cohesiveness in channel perimeter materials is less significant because the maximum velocity thread follows the centre of the channel. In fact, the channel banks are commonly vegetated right down to the bed.

3. Cross-section channel forms and streamflow characteristics

Channel features are produced, modified and moved about valley floors by the scour of bank and bed. This usually occurs at higher flows between those of average discharge and those nearly filling the channel (Wolman 1955, 1959; Harvey 1975; Pickup & Warner 1976). The recurrence interval of such effective flows can be highly variable depending on annual flow regime (Harvey 1969). Commonly it lies within a range of 5 times a year to once in 3 years. With coarse bed material the channel floor is relatively immobile until flows become competent at near-bankfull stage and a constant cross-section is maintained up to this level. In consequence the width, depth, velocity and resistance exponents at cross-sections along these channels simply describe the change in shape of the water prism as stage rises. Such assumptions, involving the rigidity of cross-section shape at lower than bankfull stages are unlikely to be valid for unconfined sand-bed channels, such as those in the semi-arid areas of south-west USA because of the greater mobility of the bed.

THE MORPHOLOGICAL CHARACTERISTICS MEASURED

Each reach was surveyed using a KERN GK1 C quickset level and metric staff (Fig 13.4), the numerous surveyed points (Table 13.1) representing the elevation of the bed and banks of the channel. The density of survey points depended on the size of the river, with cross-sections spaced about one channel width apart on the banks. Associations of these points in the cross-channel and downvalley directions provided the cross-sections and bankfull plan-form respectively. In addition long profiles of the thalweg centreline, and particular features such as point bars, medial bars, changes in bank material or vegetation and valleyside bluffs were also plotted. Field survey yielded elevations directly and distance to and between points tacheometrically, which were then employed in producing a basic 1:200 scale map. The data and digitised coordinates of survey points were then used by a computer to produce calculated cross-section parameters and to replot the map for inspection at any desired scale. 1:200 base maps were selected to ensure that the 'plottable error' (Pugh 1975) was insignificant, whilst replotted 1:500 maps proved to be the most convenient for measuring the morphological variables. A wide range of cross-section shapes were observed (Fig 13.5).

Figure 13.4   Field survey of a meander bend on the Black Water, Galloway

In all cases, the morphological form to be described is the bankfull cross-section and the planform. Once the bankfull limits of individual sections are established, the bankfull planforms follows immediately as the outline joining these points. Few of the difficulties in bankfull definition described by Riley (1972) were encountered. However, the inequality in elevation of opposite banks posed some problems. In the present cases this inequality occurs mainly where active migration is taking place towards a steep bank, leaving opposite a low point bar which grades up to floodplain level. For present purposes bankfull level is taken as that of the lower bank since relatively voluminous lateral spillage of water would take place at this level, with an associated abrupt increase in stream width. This definition was found to be appropriate for all except very asymmetric cross-sections, for which well-marked sedimentological and vegetational boundaries on the point bar provided adequate alternative field criteria.

2.   Cross-section dimensions and shape

Five variables were defined to describe the cross-section dimensions (Fig 13.6A). Cross-section shape is defined by an index of asymmetry (AS). The cross-section area is divided vertically at the midpoint of width and the index is derived as the ratio of the larger sub-area to the smaller one (Fig 13.6B). Perfectly symmetrical cross-sections produce an AS index of 1.0. When used for correlation with cross-section dimensions and pattern parameters, the index is necessarily positive. However, when compared to the direction of channel curvature, it may assume positive values where asymmetric sections are deeper towards

Bends in confined streams

the left bank (looking downstream) and negative where asymmetric sections are deeper towards the right bank. Indices describing the cross-sections in Fig 13.5 are listed in Table 13.2.

3. Plan pattern at cross-section level

In describing channel pattern at a given cross-section, two measures were defined as the angle of turn (AOT) and the direction change (DC) (Figs 13.6C and D). These indices are obtained from a straight line joining a point on a given cross-section to a corresponding point on the cross-sections upstream and downstream. An alternative measure of channel pattern at a cross-section was also computed as an arc of a circle fitted geometrically to the three points on the cross-sections and its radius of curvature (RC) estimated. Direction change and radius of curvature were calculated for two points on each cross-section. Points linking the assumed thalweg or points of deepest water are differentiated by subscript "T" compared with subscript "C" which denotes that the centre points of each section were used. As these two indices are complementary, only one, direction change (DC) is used in the following discussion.

4. Plan-form dimensions and scale at reach level

Meander wavelength could be calculated when reaches as a whole were considered. A series of discrete steps of equal length were calculated from the centreline data. The step length was scaled to the size of the river and standardised at half a channel width to the nearest metre. The orientation of each step length is expressed as a deviation from the mean direction of the reach as a whole and curvature is estimated by subtracting the directions of two adjacent steps. Curvature towards the right bank was represented by negative direction change angles and curvature to the left by positive angles. Once individual bend arcs were distinguished, according to consistency and change of sign, various wavelength estimates in metres were obtained. These fall into three groups. Firstly, along the channel, λ_{C1} is the average reach value of a number of wavelengths, each defined as the length along the channel through two adjacent bend arcs of different curvature sign (Fig 13.7A). Another estimate λ_{C2} is simply twice the average bend length along the channel. Secondly, between inflections λ_{S1} is the straight-line distance between the inflections bounding two bend arcs of different curvature (Fig 13.7B). An average value was found for a series of wavelengths corresponding to the same succession of bends as λ_{C1}. Another measure λ_{S2} was calculated as twice the average straight line distance between the inflections bounding individual bend arcs. Thirdly, a group of indices were calculated to express the scale of bends. First, the maximum perpendicular distance from the line between bed inflections to the channel was calculated (Fig 13.7C) as another expression for 'amplitude' and is termed 'arc height' (AH), as suggested by Brice (1964). Another perpendicular distance (CH), located at the midpoint of the line joining the inflections was also

Table 13.2 Indices describing typical cross-section shapes. The letter code in the first column links these indices with the actual cross-section shapes in Figure 13.5 and with the locations shown in Figures 13.2 and 13.3

Example (in Fig 13.5)	Reach number	Section number	Width (W)	Depth (D)	Area (AR)	Wetted perimeter (WP)	Hydraulic radius (HR)	Asymmetry (AS)	Width/Depth (W/D)
		m	m	m	m³	m	m	dimensionless	
A	R1	49	7.72	0.27	2.75	10.54	0.26	5.69	28.72
B	R11	31	7.37	0.44	4.24	10.35	0.41	1.26	14.69
C	R2	18	8.47	0.42	3.67	9.11	0.40	1.27	11.03
D	R5	25	14.21	0.66	10.19	16.16	0.63	1.28	21.41
E	R3	54	11.48	0.50	5.84	11.92	0.49	1.15	24.28
F	R7	26	8.17	0.79	6.49	8.85	0.73	1.24	10.10
G	R6	15	9.46	0.82	7.87	10.33	0.76	1.02	10.55
H	R11	25	9.55	0.81	7.78	10.73	0.73	1.22	29.45
I	R4	21	17.06	0.23	3.95	17.23	0.23	1.11	108.47
J	R10	4	2.61	0.05	0.16	3.73	0.04	5.57	45.86
K	R4	89	6.83	0.42	3.04	7.55	0.40	1.05	16.06

Bends in confined streams

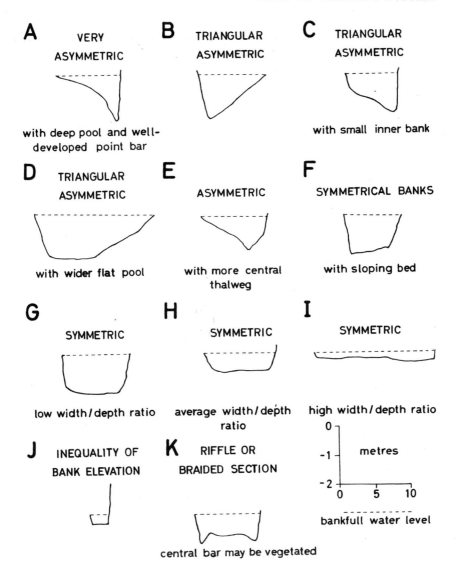

Fig 13.5 A representative range of typical cross-section shapes. Location of the sections is shown on Figs 13.2 and 13.3 and indices describing the shapes are listed in Table 13.2. The derivation of these indices is described in Fig 13.6.

measured. Two circular arcs were fitted to both AH and CH, passing through the extremities of the bends (Figs 13.7C). The second lateral bend index (ABR) is represented by the average radius of these two arcs.

Bends in confined streams

Cross section scale variables

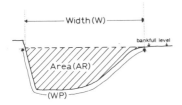

WP = Wetted perimeter

Hydraulic radius (HR) = AR ÷ WP

Mean depth (D) = AR ÷ W

A

Cross section shape variable
The Index of Asymmetry (AS)

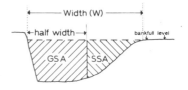

GSA = Greater sub-area

SSA = Smaller sub-area

Index of Asymmetry (AS) = GSA ÷ SSA

B

Direction Change and Angle of Turn between three consecutive sections

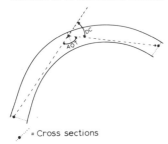

⋯ = Cross sections

• = Thalweg point within each section

↗ = Assumed flow path

DC = Direction Change

AOT = Angle of Turn

C

Radius of Curvature of a circular arc fitted to three consecutive sections

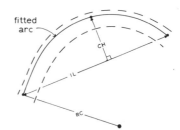

IL = Distance between upstream and downstream sections
CH = Perpendicular height to central section
RC = Radius of Curvature

$$RC = \frac{IL^2}{8CH} + \frac{CH}{2}$$

D

<u>Figure 13.6</u> Definition of cross-section dimensions and variables representing channel curvature at individual cross-sections

5. Sinuosity of plan-form shape at reach level

Stream patterns have been described qualitatively as being straight or meandering, with a transitional stage in between, meandering rivers being regular, irregular or tortuous (Schumm 1963). Three indices are used to express such sinuosity, for both the thalweg and for centreline paths. First, and most frequently, the line joining the ends of the path is assumed to follow the valley direction and sinuosity is defined as the ratio of the distance along the channel to this distance (P_C and P_T) (Fig 13.7D)

Occasionally the straight line distance may not coincide with downvalley direction and gradient, especially if the plan form is confined within valley bends, as in 'misfit' streams (Dury 1969) or where the meander belt itself shifts within a straight valley (Fig 13.2) or over a plain. Consequently, the sinuosity P_I was estimated as the ratio of channel length to the summed length of all the straight lines joining the inflections, the 'meander belt axis' (Brice 1964) (Fig 13.7D). Finally, the variance of the direction series fitted to the centre and thalweg lines has been shown to be closely, but non-linearly, related to sinuosity (P) (Ferguson 1977). This relationship was tested for these smaller rivers and the direction variances (DV_C and DV_T) were used as the third indicator of channel shape.

ANALYSIS AND DISCUSSION OF RESULTS

1. Inter-relationship at the within-reach level

a) Cross-section dimensions and asymmetry. The variables which are most highly correlated represent dimensions which are inevitably related with the depth (D), the hydraulic radius (HR) being always more closely correlated than the width (W) and wetted perimeter (WP). Correlations of wetted perimeter and hydraulic radius with each other and with depth, width, area (AR) and asymmetry (AS) naturally mirror the corresponding relationships involving width and depth. Due to these inter-correlations, little information is lost if hydraulic geometry at the within-reach level is considered in terms of the 'traditional' variables of width and depth.

There is little correlation between width and depth, ranging from $r = -0.47$ for the River Rede to $r = 0.53$ for Bowmont Water, due partly to the variability of both widths and depths, but particularly depths. For each reach the coefficient of variation of depth is usually at least twice that of width, with the actual amounts assuming values in the ranges 15.86 - 32.28 per cent and 16.96 - 83.7 per cent, respectively. In addition, the variation in width is not matched by variation in depth in the same direction or degree. An increase in width from one section to the next may be accompanied by an increase or decrease in depth to a comparable, lesser or greater, degree. Lack of coincidence of change in dimensions also reflects the fact that two sections of the same width can have markedly different shapes, depending on their location in the planform. This variability in itself would tend to suggest that adjustment of cross-sections to the dominant flow varies differently and frequently as local environment changes or that individual sections require different discharges to overcome local conditions. However, the variation of cross-section depth and shape with a relatively constant width will also be influenced by larger-scale underlying pseudo-cyclic trends related to the channel curvature in plan and the pool-riffle sequence.

The scatter of the weak relationships between width and depth has been examined as large residuals from the

least-squares regression and then considered in terms of
the geographical location of that point. Examination of
the residuals greater than one standard error for the
estimated depth value from width supported a separation of
the scatter into random environmental constraints and more
systematic variations in cross-section shape. Most
commonly occurring are negative residuals from the width
v. depth relationships, implying a lack of response to
adjust depth. Closer inspection indicated that groups of
these residuals occur at three specific geographical
locations (Milne *in preparation*). First, significant
negative residuals can be found along the parts of bends
which impinge on valley-side bluffs of unconsolidated
valley-fill. The confining effect of either the valley-
fill bluffs or the bedrock walls varies, depending on the
angle at which the channel makes contact with the valley-
side (Fig 13.8). Greater angles tend to induce tighter
turns, even where valley-fill is actively undercut. High
sediment production may characterise such locations
(Lewin & Brindle 1977) and boulder supply from the sheer,
unstable bluffs may be sufficient for points of channel
contact with the valley side to be relatively immobile even
in unconsolidated materials. The widths corresponding to
these depths are no greater or less than average, indicating
that the channel is too shallow for that width. It seems
that the channel is unable to adjust its depth because of
the high supply of coarse material at these points.
Isolated negative residuals elsewhere possibly indicate
similar local concentrations of immobile materials or lack
of competence due to locally low gradients. The second
important location of large negative residuals is cross-
sections located on riffles due essentially to a lack of
depth adjustment. In other cases, however, riffles in
straight or low sinuosity reaches have developed greater
widths due to divergent flow around bars (Richards 1976).
The coarse nature of riffles is well-known (Leopold *et al*
1964) and some of these sites may coincide with loci of
residual bed material. An estimate of residual material
was made by comparing the distribution of bed-material size,
measured along the Kingledoors Burn, with the maximum size
of material the stream was competent to move, as derived
from the relationship established by Baker & Ritter (1977).
Concentration of residual material below a valley-side
bluff, and, to a decreasing degree, downstream, firmly
established this location as an area of coarse, immobile
sediment supply. Thirdly, significant negative residuals
are found in deep asymmetric pools on well-developed bends,
often adjacent to the riffle sections. Depending on the
angle at which the stream current impinges on such points,
tight turns may again be induced. However, the cohesiveness
of such materials favours deeper scour, and markedly
asymmetrical pool and point bar cross-sections develop.
The greater width relates to the migratory nature of these
sections which also occur at distinctive locations. They
occur in bends with a relatively small radius of curvature
developed in fine channel fills along the outer bank
leaving gravel point bars which grade gently up to floodplain
level. The high silt-clay content of these bank deposits

ensures the outer banks are steep and the sections asymmetric. This situation contrasts with that of sand beds in lowland rivers where depth is more easily adjusted and cross-sections remain narrow and deep as migration proceeds, and is expressed in relationships between channel sediment, width-depth ratio and sinuosity (Schumm 1963). Floodplain paleofeatures may also be important at these points. Bank collapse is particularly likely where paleo-channel fills meet the present bank almost perpendicularly due to the extra moisture which collects here reducing bank strength (Knighton 1973).

Location of significant positive residuals in the width v. depth relation indicated the importance of planform to cross-section shape because they are represented by pools that are located in the longitudinal sequence but which occupy gently curved or straight reaches (Milne *in preparation*). While the maximum depth of these pools is equivalent to that of the migrating pools in tightly curved reaches, the symmetry imparted by lack of curvature means their area and therefore average depth is considerably greater. As widths are usually no greater than average this indicates an ability to scour to a greater depth, which may indicate locally erodible bed materials, perhaps a less erodible bank environment, or the location of locally high flows in the pool-riffle sequence may be important (Keller 1971). Other isolated positive residuals seem to be random sections where width adjustment is restricted rather than an over-adjustment of depth.

Correlations between depth and area are greater than $\underline{r}$ = 0.54 and most are between $\underline{r}$ = 0.8 and $\underline{r}$ = 0.96 suggesting that much of the variation in area is explained by changes in depth. A given increase in depth will naturally cause a greater percentage increase in area in a wide, shallow section than in a deeper section of the same width. Hence in the regression of depth v. area, all negative residuals correspond to sections narrower than average and conversely all positive residuals occur at sections wider than usual. The latter group is that already singled out as the migrating pool sections.

Any relationship between asymmetry and width, depth or area is, at best, erratic with few coefficients exceeding $\underline{r}$ = 0.5 and the majority are less than $\underline{r}$ = 0.3. One reason for the low correlations in these cases is that asymmetry is only pronounced in pool sections on migrating bends where asymmetry values reach at least 5.0. Elsewhere values rarely exceed 1.5.

b) Cross-section variables v. local direction change and radius of curvature. Commonly tighter bends, by imposing a greater transverse secondary circulation, are associated with greater asymmetry. Strikingly, this close association between sinuosity and cross-section shape is not yet established within the upland reaches of the streams included in the present study, with correlations rarely exceeding $\underline{r}$ = 0.35. However, residuals from the approximate relationship between direction change and

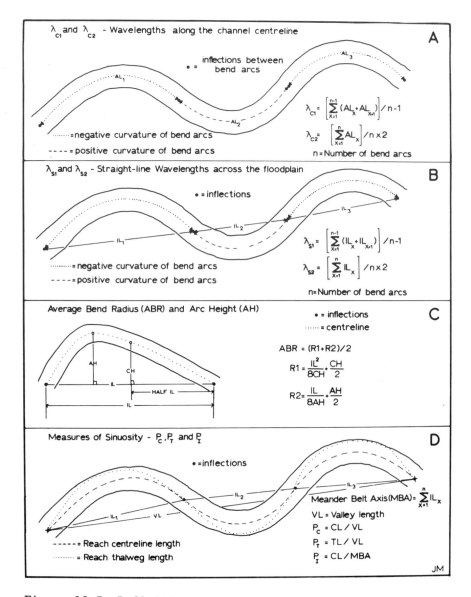

Figure 13.7 Definition of variables representing meander planform dimensions and shape at a reach level.

asymmetry are noteworthy since they may indicate the possible factors which are restricting adjustment. For instance, large positive residuals deviate from the general trend where, with small values of asymmetry, direction change is less than usual. The localities involved are bluffs where high sediment supply has straightened the flow path by shifting the thalweg towards the centre of the channel. Negative residuals tend to be associated with sharp turns where the input of material inhibits scour.

Figure 13.8   Erosion scar in the Kingledoors valley-fill, showing the accumulation of coarse material at the scar base ready for supply to the stream channel.

c) *Comparison of within-reach results at an inter-reach level.* Within a given reach the magnitude and direction of morphological relationships are best expressed by average width-depth ratios (Table 13.1) and sinuosity, preferably P_I. Actual values of sinuosity, P_I, are not high. They barely reach the 'transitional' stage defined by Schumm (1963), whilst none of the greater values of P_C and P_T qualifies as 'tortuous'. The migratory reaches are the most sinuous, such as Kingledoors and Stanhope Burns and both the Black Water reaches. The lowest values represent relatively straight reaches, separated by sharp turns, as in the River Ure reach. Of intermediate wiggleness are Polmaddy Burn and Snaizeholme Beck. Paralleling this decrease in sinuosity is also a decrease in width-depth ratio, the migrating channels being wider and shallower than the relatively straight reaches which are bounded on both sides by steep banks in fine materials. In contrast Schumm (1963) established a negative relation between width-depth ratio and sinuosity, implying that the narrower and deeper a channel, the more sinuous it will be. However, Schumm studied only large lowland rivers flowing over much gentler gradients, which had virtually no material coarser than gravel. Nonetheless, the general contrast in the relationships observed is probably related to the considerable thickness of fine floodplain material along the gentler gradients of lowland which allows

Table 13.3 Average reach values for cross-section, meander scale, and shape variables.

Cross-section

	R1	R2	R3	R4	R5	R6	R7	R8	R9	R10	R11
W	5.94	7.45	8.71	8.30	14.11	10.97	8.42	9.04	8.06	15.51	2.54
D	0.27	0.36	0.49	0.45	0.80	0.79	0.58	0.78	0.61	0.63	0.09
AR	1.74	2.83	4.37	3.89	11.75	8.81	5.11	7.43	4.92	10.60	0.27
WP	6.48	8.10	9.41	8.96	14.99	11.97	9.30	10.15	9.14	17.00	3.28
HR	0.25	0.34	0.47	0.42	0.77	0.74	0.55	0.72	0.55	0.61	0.08
AS	2.05	1.59	1.66	1.61	1.24	1.24	1.32	1.31	1.80	1.73	2.33

Meander scale

	R1	R2	R3	R4	R5	R6	R7	R8	R9	R10	R11
$\lambda C1$	87.9	95.2	151.7	151.0	172.0	140.0	123.4	124.1	103.4	185.0	51.0
$\lambda C2$	89.3	95.1	153.8	158.4	180.0	140.0	127.4	128.3	98.0	185.0	51.0
$\lambda S1$	63.0	83.1	125.5	133.1	160.1	123.2	116.3	103.5	87.2	182.3	44.9
$\lambda S2$	72.5	85.2	135.3	139.4	172.6	136.3	123.2	111.3	87.3	182.3	44.5
ABR	68.9	61.4	79.9	69.8	177.6	177.9	508.6	226.4	84.7	361.2	15.1
AH	9.66	7.85	12.79	13.82	9.48	4.53	5.05	9.39	7.04	3.30	5.05

Sinuosity

	R1	R2	R3	R4	R5	R6	R7	R8	R9	R10	R11
P_C	1.63	1.23	1.27	1.24	1.34	1.06	1.09	1.69	1.87	1.01	1.14
P_T^C	1.74	1.27	1.31	1.28	1.39	1.10	1.10	1.72	1.94	1.06	1.18
P_I	1.23	1.12	1.14	1.14	1.04	1.03	1.03	1.15	1.15	1.02	1.15
DV_I^C	0.92	0.42	0.47	0.45	0.55	0.11	0.17	0.94	1.14	0.02	0.26
DV_T^C	1.05	0.47	0.54	0.51	0.62	0.18	0.19	0.98	1.20	0.12	0.32

Planform shape

	R1	R2	R3	R4	R5	R6	R7	R8	R9	R10	R11
DC_T	0.18	0.19	0.19	0.17	0.13	0.10	0.09	0.14	0.17	0.07	0.16
DC_T	0.26	0.31	0.32	0.28	0.25	0.30	0.18	0.31	0.27	0.37	0.21
RC_C	261.1	125.5	332.4	238.8	612.6	10464	312.7	367.2	423.8	403.7	58.3
RC_T^C	131.5	112.1	138.8	107.8	142.0	79.1	105.8	109.8	134.2	44.2	43.0

both banks to remain steep. The easier depth adjustment of sandy beds means that sections can remain narrow and deep while migration increases sinuosity. In contrast in the upland environment the typical migrating section is very asymmetric, with a deep pool and well-developed point bar due to the coarseness of the bed material. Upland straight reaches can be narrow and deep, such as Carsphairn Lane, or wide and shallow like Bowmont Water. However, if migration is initiated in a narrow deep reach, the rapid building of a point bar, once migration has proceeded sufficiently far from the original steep inner bank, creates a wide and shallow cross-section.

The straight reaches of <u>low</u> width-depth ratio are less variable than migrating bends in width, depth, area and asymmetry. Meaningful relationships emerge more clearly. Morphological relationships with asymmetry are most clearly established at sharp migrating bends at which asymmetry is clearly developed.

2. <u>Relationships at reach level</u>

 a) *Inter-relationships among scale and shape variables*. By averaging the indices for each reach (Table 13.3) the

Bends in confined streams

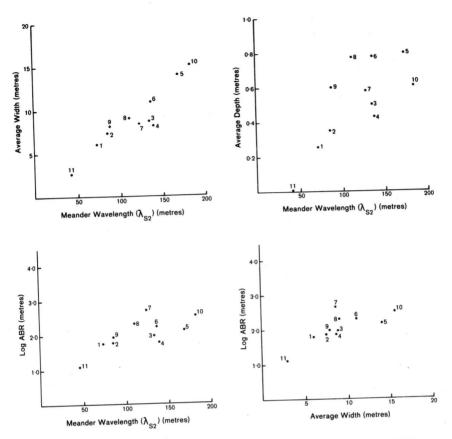

Figure 13.9   Inter-relationships between average cross-section and planform dimensions

sample size is vastly reduced to $\underline{N}$ = 11. However, uncertainties due to scatter in the data are also greatly reduced. Inter-correlations between the section dimensions are all highly significant. Virtually functional relationships are now established for wetted perimeter and hydraulic radius with width and depth, but the width v. depth relationship is not exact ($\underline{r}$ = 0.79). Average width correlates slightly more closely with area than depth. The average asymmetry index consistently produces a significant but negative relationship with all cross-section dimensions. This inverse correlation reflects that, for the reaches studied, the smaller streams include the reaches where migration and hence asymmetry is most pronounced. In addition the smaller channels include the tighter bends sufficient to produce asymmetry values of 2.0 or more.

Similarly the averages for scale variables representing wavelength are virtually functional relationships ($\underline{r}$ > 0.97). Strikingly the straight-line wavelengths λ_{S1} and λ_{S2} are consistent with the broadly established pattern, by falling in the 10-14 times width. The average arc

Bends in confined streams

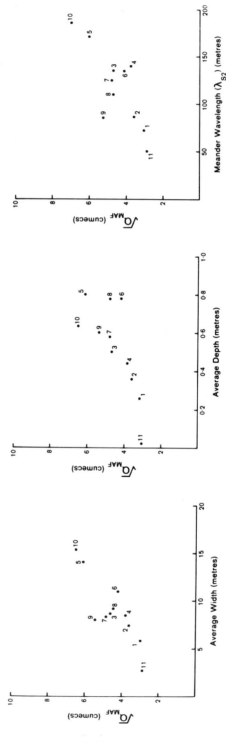

Figure 13.10 Relationships of flow magnitude with average cross-section and planform dimensions

height (AH) shows little relationship with wavelength estimates and average bend radius, which is consistent with the view that amplitude of meander is "... determined by erosion characteristics of the stream banks and by other local factors than by simple hydraulic principles" (Leopold & Wolman 1960).

The relationship between average values for planform shape indices, with the exception of the P_I index, are virtually functional. Both the centreline and thalweg direction variances and sinuosities (DV_C, DV_T, P_C, P_T) fit excellently with the non-linear relationship between sinuosity and direction variance proposed by Ferguson (1977).

Average values also bring out the generally long, low waveform of these bends in upland valleys. Ratios of arc height to the distance between bend inflections are well below 0.2, except for Kingledoors and Halter Burns. Comparable ratios for the larger freely meandering rivers would exceed 1 (eg. Leopold & Wolman 1860; Carlston 1965; Chitale 1970). Migration in unconfined channels is typically an increase in amplitude or 'loop expansion' (Daniel 1971). In contrast, streams in confined upland valleys tend to translate their waveforms in a downvalley rather than in an across-valley direction. Chitale (1970) suggests a physical mechanism for these migratory differences. In narrow, deep sections, the maximum shear around the bend is observed at the apex, causing a direct increase on amplitude and increasing sinuosity. However, in wider, shallower streams, flowing over coarse bed materials, weaker transverse flow and higher velocity promote maximum erosion along the bank downstream from the apex. In consequence, migration is a downvalley translocation, with the amplitude of the meander wave remaining low.

b) Relationships between cross-section scale and meander scale and shape. The cross-section dimensions correlate significantly with all the wavelength estimates, especially width, consistent with the link of meander scale with flow magnitude as expressed indirectly by channel width (Fig 13.9). In all cases width correlates more closely with wavelength rather than with other estimates of plan-form scale, followed by area, depth, and finally asymmetry. All the cross section v. meander scale relationships are positive, except for those with asymmetry. It appears that the wider, deeper channels have longer, larger bends, and, for the cases studied here, that it is the smaller-scale reaches which exhibit tighter bends and high asymmetry values.

Correlations between logarithmically transformed average bend radius (ABR) and the cross-section variables may exceed r = 0.7. However, average reach values for radii of curvature are highly variable and no correlations with cross-section data emerge. The only significant correlation with the centreline data indicated that the greater the direction change, and therefore the sharper the turn, the smaller the area was. This apparent relationship is probably simply an expression of the make-up of the present reaches studied, in which the smaller streams chanced to be those which were migrating and which therefore possessed sharper turns.

The insignificance of an association between cross-section dimensions and arc height probably supports the view that lateral meander development is related more with local environmental conditions than with hydraulic factors or width.

c) Relationships between scale variables and discharge. Of the stream studied, only Snaizeholme Beck is reliably gauged. Indeed, it may be simply the availability of discharge records for larger streams which has biased the bulk of previous work toward the consideration of larger rivers only. Discharge was estimated for the present cases by the computation of the 'mean annual flood' which has an estimated recurrence interval of 2.33 years (Dury 1969). This measure (Q_{MAF}) was derived indirectly (Table 13.1) using drainage basin characteristics and climatic variables in a multiple regression equation (Natural Environment Research Council 1975). The square root of the discharge estimates derived ($\sqrt{Q_{MAF}}$) was found to correlate more closely with all the scale variables (Fig 13.10). Of these, the correlation with width is most clearly expressed reflecting its control on meander scale, although the relationship with area is also close.

CONCLUSIONS

Eleven upland gravel-bed channels have been studied in order to test the breadth of applicability of generalisations derived from the previous, numerous studies on much larger, freely meandering rivers of sandy lowlands. The sand beds of such lowland rivers shift over a range of flows well below bankfull. An important, contrasted characteristic of cross-sections in the coarse bed material of confined upland channels is the relative stability of the overall cross-section shape at flows less than the dominant discharge. Although the cross-section shape is comparatively stable, it is also much more irregular in upland channels, as shown by the large coefficients of variation, particularly for depth, area, wavelength and radius of curvature. This inherent variability makes it difficult to discern significant relationships in the case of a single stream. However, by taking a very large number of measurements from a geographically varied range of cases, and by calculating their average values, upland channels are found to conform to general rules regarding channel scale. Wider channels are also deeper and develop larger bends and wavelengths. River size is related to the square root of flow magnitude, with the link between discharge and plan form being accomplished by width. Lack of correlation of the scale variables with planform shape, however, reflects the impedance to the rapid modification of upland channel sizes and shapes. The generally low arc height, sinuosity and direction variance values, and the small variation in average direction-change and angle-of-turn reflect the character of the waveforms in this environment compared to the much more tortuous lowland rivers. This lack of amplitude development is probably due to combination of the confining effects of valley walls and the nature of channel migration. Migrating cross-sections in the upland environment tend to be wider and shallower than their lowland counter-parts which inhibits well-developed transverse flow and very tight turns and which, in combination with higher velocity promoted by steeper valley gradients, suggests that these channels tend to translate the same low wave form downvalley rather than to increase in amplitude.

ACKNOWLEDGEMENTS

This investigation was funded by a Post-graduate Research Studentship from the University of Hull, and supervised by Dr. R. I. Ferguson, lecturer in Environmental Sciences, University of Stirling. A full list of field assistants would be lengthy, and to mention Andy Naylor, Steve Foster, Andy Brayshaw, Paul Burrows and Derek Grounds in particular is to omit many others to whom I am also grateful. For photographic work I am indebted to Mr. E. Quenet and Mrs. Doreen Shanks of the Department of Geography, University of Newcastle.

REFERENCES

Baker, V.R. & Ritter, D.F., 1977. Competence of rivers to transport coarse bedload material. *Bulletin of the Geological Society of America,* 86, 975-978

Brice, J.C., 1964. Channel patterns and terraces of the Loup Rivers in Nebraska. *United States Geological Survey Professional Paper,* 422-D

Carlston, C.W., 1965. The relation of free meander geometry to stream discharge and its geomorphic implications. *American Journal of Science,* 263, 864-885

Chitale, S.W., 1970. River channel patterns. *Proceedings of the Hydraulics Division, American Society of Civil Engineers,* 96(HY1), 201-222

Dury, G.H., 1969. Relation of morphometry to runoff frequency. In: *Introduction to fluvial processes.* ed. R. J. Chorley (Methuen, London), 177-188

Ferguson, R.I., 1977. Meander sinuosity and direction variance. *Bulletin of the Geological Society of America,* 88, 212-214

Harvey, A.M., 1969. Channel capacity and the adjustment of streams to hydrologic regime. *Journal of Hydrology,* 8, 82-98

Harvey, A.M., 1975. Some aspects of the relations between channel characteristics and riffle spacing in meandering streams. *American Journal of Science,* 275, 470-478

Keller, E.A., 1971. Areal sorting of bed material: the hypothesis of velocity reverasal. *Bulletin of the Geological Society of America,* 82, 279-280

Keller, E.A., 1972. Development of alluvial stream channels: a five-stage model. *Bulletin of the Geological Society of America*, 83, 1531-1536

Knighton, A.D., 1973. River bank erosion in relation to streamflow conditions, River Bollin-Dean, Cheshire. *East Midland Geographer*, 5, 416-426

Leopold, L.B., 1973. River channel change through time: an example. *Bulletin of the Geological Society of America*, 84, 1845-1860

Leopold, L.B. & Wolman, M.G., 1957. River channel patterns; braided meandering and straight. *United States Geological Survey Professional Paper*, 282-B

Leopold, L.B. & Wolman, M.G., 1960. River meanders. *Bulletin of the Geological Society of America*, 71, 769-794

Leopold, L.B., Wolman, M.G. & Miller, J.P., 1964. *Fluvial processes in geomorphology* (Freeman, San Francisco and London)

Lewin, J. & Brindle, B.J., 1977. Confined meanders. In: *River channel changes*, ed. K. J. Gregory, (Wiley, Chichester), 221-232

Natural Environment Research Council, 1975. *Flood studies report*, Volume 1. Hydrological studies; Volume v. Maps. (Her Majesty's Stationery Office, London)

Pickup, G. & Warner, R.F., 1976. Effects of hydrologic regime on magnitude and frequency of dominant discharge. *Journal of Hydrology*, 29, 51-75

Price, R.J., 1963. The glaciation of a part of Pebbleshire. *Transactions of the Edinburgh Geological Society*, 19, 325

Pugh, J.C., 1975. *Surveying for field scientists.* (Methuen, London)

Richards, K.S., 1976. Channel width and riffle-pool sequence. *Bulletin of the Geological Society of America*, 87, 883-890

Riley, S.J., 1972. A comparison of morphometric measures of bankfull. *Journal of Hydrology*, 17, 23-31

Schumm, S.A., 1963. Sinuosity of alluvial rivers on the Great Plains. *Bulletin of the Geological Society of America*, 74, 1089-1100

Schumm, S.A., 1967. Meander wavelength of alluvial rivers. *Science*, 157, 1549-1550

Simons, D.B. & Richardson, E.V., 1960. The effect of bed roughness on depth-discharge relations in alluvial channels. *United States Geological Survey Water-Supply Paper*, 1498-E

Wilcock, D.N., 1967. Coarse bedload as a factor determining bed shape. *Publication of the International Association of Scientific Hydrology*, 75, 143-150

Wolman, M.G., 1955. The natural channel of Brandywine Creek. *United States Geological Survey Professional Paper*, 271

Wolman, M.G., 1959. Factors influencing erosion of a cohesive river bank. *American Journal of Science*, 257, 204-216

CHAPTER 14   THE OVERALL SHAPE OF LONGITUDINAL PROFILES OF STREAMS

D. A. Wheeler[*]

INTRODUCTION

Compared with other fluvial phenomena, the longitudinal profile of streams is the least transient expression of the effect of running water on the earth's land surface. In consequence, the effects of past conditions and events are more clearly preserved in their particular shapes than in the more ephemeral fluvial features or short-term hydrological events. For instance, many steps in longitudinal profiles may be attributed to the work of long-departed glaciers and others might appear to be associated with river diversions (Lewis 1945). Indeed, the significance of breaks in gradient ".. is vital in the reconstruction of former stages in the river's history" (Brown 1952). Despite the relative immutability of the legacies of such past events, less discernible changes in other environmental conditions, and the infinite variety in the outcrops of solid geology, there is a marked and widespread tendency to treat the shape of a stream's longitudinal profile as an idealised form. Some assumptions about interdependence or even equilibrium between the factors and forces involved are usually adopted and it is widely believed that rivers tend to work progressively towards an increasingly smooth profile shape. For example W. W. Rubey felt convinced "... that certain broad principles must underlie and explain the systematic smoothness of these curves" (Shulits 1941). J. H. Mackin's (1948) much-quoted definition of the 'graded' stream is that of a delicately adjusted gradient which, over a period of years under prevailing conditions, provides just the velocity required to remove all waste supplied to the drainage basin. The contrasted view of Leopold & Langbein (1964), with the long profile as an expression of energy distribution, supposes that a quasi-equilibrium is maintained between the opposed tendencies of entropy and minimum work.

It is striking that interpretations of the long profile, the most obvious, inevitable and least changing expression of fluvial processes, should be coloured by so many shades of meaning. However, it is clear that the value of the stream profile as the long-term summary of the effectiveness of fluvial processes and as an historical record has been obscured amid the arguments surrounding the extrapolation of segments of stream profiles, the attendant risks of

---

[*] Department of Geography, University of Keele

assuming constancy of sea-level for protracted periods, and the widespread assumptions about the overall smoothness and regularity of its idealised shape. The present account is based on the supposition that there could be many advantages simply in generalising objectively and numerically the actual shape of real stream profiles without partiality to any of the pre-conceptions which abound.

METHODS AND DATA COLLECTION

Until recently, studies of actual long profiles of streams usually considered no more than six cases, and commonly an individual stream or a major reach of a stream were surveyed in detail. For example, field surveys by K. C. Boswell (1946) on the River Test, and E. H. Brown (1952) on the Ystwyth followed the work of O. T. Jones (1924) on the Towy, noting his insistence on field survey. Clearly this approach is essential in elucidating the critical details of a specific case. However, if the variabilities of geology and geomorphological history are to be encompassed in broader generalisations, the uppermost criterion becomes that of a very large number of geographically diverse cases. This criterion can only readily be met by extracting the necessary detail from highly accurately contoured maps from an area within which map cover is reasonably complete. Thus the geological diversity of Britain, together with the degree of map coverage at 1:25 000 scale, favoured the examination of stream longitudinal profiles within the British Isles as the basis for the present enquiry. In all the 115 rivers studied averaged 80.5 km in length, with a mean fall of 366 m summarising the 'stream relief'. Although an even coverage of streams in England, Wales and the Southern Uplands of Scotland was possible, map coverage at the 1:25 000 scale is not complete in the Grampians (Fig 14.2). Although several ways of summarising the data quantitatively have been examined (Wheeler *in preparation*), attention here is focussed on stream profile concavity, partly because it is assumed to be "... so common as to be almost universal" (Rubey 1933) and partly because varying degrees of overall concavity can be measured objectively with little complication (Langbein 1964). However, Langbein's index of concavity was originally devised to describe profiles of a much greater size-order and different character such as the River Nile.

The size at which the long profiles were plotted was standardized, so that all shapes could be compared regardless of river length. The dimensions of the standardised plot were controlled by practical considerations. In the present instance, the full width of computer graph plotter output allowed the drawing of profile length at 61 cm and, for convenience in filing, each profile was drawn to a height of 23 cm. Two lines were then drawn on the profile plots, one joining the headwaters point to the estuary or tributary junction, with the other parallel to it, but being as far from the first line as possible whilst still being in contact with the profile outline. The index of concavity used is that of the length of the perpendicular

Figure 14.1　A classic example of the marked change in stream channel gradient typical of many rivers in upland Wales - the River Rheidol at Devil's Bridge, east of Aberystwyth.

between these two lines divided by the height of the plot. This procedure permits ready comparison of concavity estimates made using plots of a different size. Thus, Langbein's measure, originally used on the profiles of a small number of continental-sized rivers, is modified slightly to represent the smaller but more diverse forms of stream profiles in the British Isles. The measures of profile concavity obtained (Table 14.1) averaged 0.43 units and varied from as little as 0.089, the Colne in Hertfordshire, to as high as 0.70 for the River Derwent in Cumbria. However, within most major regions of the British Isles, a range of concavity values is revealed.

THE MAIN VARIABLE CORRELATING WITH LONG PROFILE CONCAVITY

W. Langbein (1964) found that, for a small sample of rivers drawn from semi-arid environments, long profile concavity correlates with stream length. This relationship was not found to obtain in the case of British rivers ($r = 0.11$, $N = 115$). Indeed, of a number of variables including stream relief, mean discharge, basin area and mean gradient, as well as stream length, only stream relief correlated notably with profile concavity, $r = 0.58$ (Fig 14.3). This single variable thus accounts for 34 per cent of the variability in concavity values. To examine the significance of this association, climatological and hydrological data were extracted from the Surface Water Yearbook, and a matrix of correlations between drainage area, station elevation, basin relief, mean annual rainfall, mean annual runoff and mean discharge were established (Table 14.2). The inevitable dependence of

Shape of longitudinal profiles

Table 14.1   Index code numbers for stream profiles and the calculated concavity index, CI

No.		CI	No.		CI	No.		CI
1.	Adur	.50	40.	Esk	.54	78.	Seven	.42
2.	Air	.54	41.	Exe	.41	79.	Severn	.65
3.	Aln	.44	42.	Exe	.34	80.	Stinchar	.45
4.	Annan	.58	43.	Fal	.35	81.	Stour	.49
5.	Arun	.52	44.	Fowey	.19	82.	Stour	.30
6.	Avon	.27	45.	Frome	.28	83.	Stour	.33
7.	Avon	.29	46.	Girvan	.44	84.	Swale	.46
8.	Avon	.37	47.	Glaslyn	.68	85.	Taff	.41
9.	Axe	.33	48.	Great Eau	.42	86.	Tamar	.39
10.	Ayr	.33	49.	Greta	.21	87.	Taw	.54
11.	Blyth	.30	50.	Halladale	.35	88.	Tawe	.46
12.	Brora	.42	51.	Hodge B.	.31	89.	Tees	.47
13.	Calder	.40	52.	Hope	.58	90.	Teifi	.53
14.	Camel	.39	53.	Irt	.72	91.	Teign	.47
15.	Clwyd	.45	54.	Irthing	.28	92.	Teme	.42
16.	Clyde	.36	55.	Irvine	.46	93.	Test	.17
17.	Colne	.09	56.	Kennet	.11	94.	Teviot	.36
18.	Conway	.45	57.	Kent	.68	95.	Thames	.26
19.	Coquet	.38	58.	Leven	.41	96.	Till	.56
20.	Cree	.49	59.	Lune	.49	97.	Torridge	.27
21.	Dart	.41	60.	Lyne	.44	98.	Towy	.48
22.	Dee	.57	61.	Mawddach	.46	99.	Trent	.46
23.	Dee	.44	62.	Medway	.48	100.	Tweed	.38
24.	Derwent	.70	63.	Mersey	.52	101.	Tyne	.39
25.	Derwent	.53	64.	Mole	.12	102.	Tyne	.49
26.	Derwent	.66	65.	Monnow	.50	103.	Ure	.47
27.	Dionard	.63	66.	Naver	.59	104.	Usk	.46
28.	Don	.52	67.	Neath	.55	105.	Waver	.54
29.	Doon	.43	68.	Nidd	.53	106.	Wear	.47
30.	Dove	.38	69.	Nith	.39	107.	Wey	.48
31.	Dove	.46	70.	Otter	.26	108.	Wharfe	.36
32.	Dovey	.61	71.	Ouse	.43	109.	Wh'adder	.23
33.	Duddon	.54	72.	Ouse	.43	110.	Wick	.33
34.	E. Cleddau	.42	73.	Parratt	.63	111.	Witham	.52
35.	Ebbw	.32	74.	Ramsdale	.13	112.	Wnion	.54
36.	Eden	.57	75.	Ribble	.48	113.	Wye	.56
37.	Ellen	.53	76.	Rother	.37	114.	Wyre	.59
38.	Elwy	.30	77.	Rye	.37	115.	Ystwyth	.32
39.	Esk	.57						

Rivers of the same name are located in Clwyd (22); Cumbria, 24, 39; Derbyshire, 25, 30; Dorset, 6, 81; Essex, 82; Galloway, 23, 40; Gloucestershire, 7; Kent, 83; Lothian, 101; Norfolk, 71; Northumberland, 102; Sussex, 72; Warwickshire, 8; Yorkshire, 26, 31, 41.

Shape of longitudinal profiles

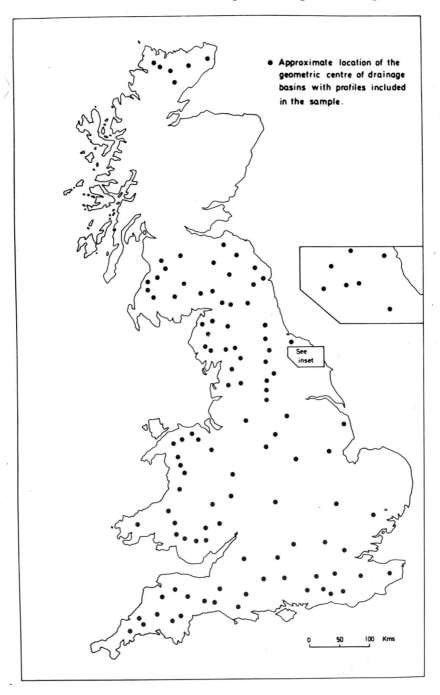

Figure 14.2 Location of streams for which longitudinal profiles are investigated

Shape of longitudinal profiles

runoff on rainfall and of discharge on drainage area is specified, but the degree of control of altitude on rainfall and runoff is established. This last relationship could imply that profile concavity may be accentuated by the greater effective precipitation at higher altitudes in an area with a humid temperate environment such as the British Isles. In such environments it seems that the rate of increase of discharge is greater for streams of greater relief, as the increase of runoff and rainfall with altitude means that more water per unit length of channel is being introduced into an upland stream than into a lowland stream.

Gilbert (1877) has previously drawn attention to the inverse relationship which exists between gradient and the volume of water. In consequence, if along a stream's course the rate of increase of discharge is rapid, then the rate of decrease of gradient will be proportionately greater. If this condition is fulfilled, then strongly upward-concave long profiles might be readily explained. The validity of this hypothesis may be examined using data from the Surface Water Yearbook. For fifteen rivers there is a sufficient downstream sequence of gauging stations to permit individual regression analyses to describe the rate of downstream change in the discharge-drainage area relationship. In all cases a significant degree of correlation was established. More relevant, however, is the significant correlation which also exists between the regression coefficient (rate of discharge increase downstream) and profile concavity ($\underline{r}$ = 0.59, $\underline{N}$ = 15). An equally informative association also exists between profile concavity and annual runoff ($\underline{r}$ = 0.54). These conclusions indicate a need to examine overall profile form in terms of the dominant hydrological characteristics of the basins in question.

POSSIBLE REASONS FOR ATYPICALLY SHAPED LONG PROFILES

Due to the large number of cases included in the present study, the slope of the regression line in Fig 14.3 defines reliably the stream relief - concavity relationship typical for rivers in the British Isles. In relation to the 'typical' case for the British Isles, certain atypical profiles can be identified as points lying away from the line of the regression equation. These can be described, in relative terms, as 'under-concave' or 'over-concave'. It is worthwhile considering some of the more atypical cases, as their features may be sufficiently pronounced to indicate other factors influencing long profile shapes.

1. 'Over-concave' profiles

There are only two streams which have markedly 'over-concave' profiles (Fig 14.3), the Yorkshire Derwent and the River Parrett. In both cases geological similarities may be detected. The rivers rise at opposite ends of the lengthy outcrop of Jurassic strata which runs in a northeast to south-west direction across Midland England. The Derwent (Fig 14.4) rises at 261 m on the eastern flanks of the North York Moors and the Parrett at 147 m on the Dorset

		1	2	3	4	5
1.	Drainage area					
2.	Station elevation	-0.36				
3.	Maximum elevation	0.19	0.43			
4.	Mean annual rainfall	-0.20	0.44	0.75		
5.	Mean discharge	0.84	-0.14	0.53	0.25	
6.	Mean annual runoff	-0.13	0.42	0.75	0.88	0.32

$N = 300$

Table 14.2 Correlation matrix of hydrological variables, demonstrating the degree to which greater quantities of surface water are present at higher altitudes in the British Isles

Heights. Thereafter both streams fall relatively rapidly to lower levels. The Derwent is down to an elevation of 25 m within 27 km of its 112 km course. The Parrett is at the same elevation within 13 km of its 64 km course. Both streams then follow courses which are largely over areas of extensive glacial and post-glacial infilling. The Derwent flows over the lowlands of the Vale of York before joining the Ouse system near Hemingborough, and, similarly, the Parrett follows a course over the Somerset Levels. If recent infilling of the greater part of the lower reaches of a river by glacial and post-glacial deposits can create such degrees of over-concavity in the long profile shape as a whole, it is important to consider other infilled areas and to note the degree of concavity. Such streams end in many of the major estuarine lowlands in England, such as the estuaries of the Severn (Fig 14.4), Mersey, Dee and Humber, and in the Wash and Solway Firth. For example, the Eden flows into the Solway Firth after crossing the drift-covered Vale of Eden. The Trent, Don, Aire and Calder all flow over lowlands at the head of the Humber. The River Witham adopts a circuitous route before entering the Wash. It is noteworthy that in each of these cases some measure of 'over-concavity' is detected in the river long profiles (Fig 14.3 and Table 14.3A). In all these areas, appreciable depths of recent deposits have been proved. For example, at Wetheral the Eden is underlain by 24 m of deposits (Hollingworth 1931). The lower Dee and Mersey are at least 15 m above earlier courses (Gresswell 1964), and bedrock beneath the Humber at Hessle is encountered at -25 m OD. At the Severn Tunnel site, bedrock is encountered at -22 m OD (Wills 1938) and the comparable figure for the Parrett is -27 m OD (Godwin 1960). Infilling is notable in other areas such as the South Wales rivers the Taff, Rhymney and Usk (Williams 1968), but since infilling is limited to only relatively short reaches of these estuaries, the effect is not sufficiently pronounced to impart any degree of over-concavity to the overall long profile shape. In all of these rivers which end in estuarine lowlands, with the exception of the Wash, the underlying bed-rock is Permo-Triassic strata.

Shape of longitudinal profiles

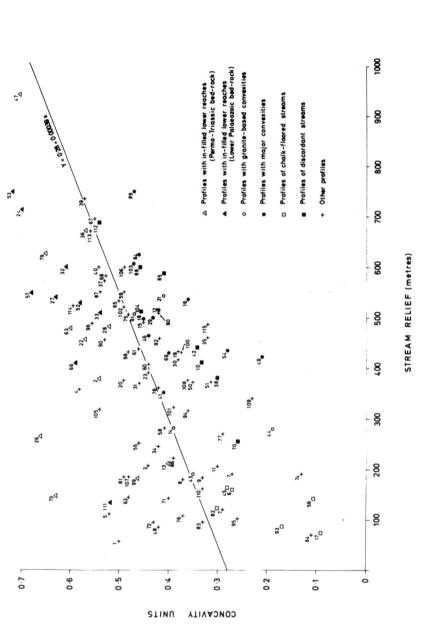

Figure 14.3 The association of stream relief with overall concavity in longitudinal stream profiles

Shape of longitudinal profiles

A. DRIFT-FILLED LOWLANDS		B. AGGRADATION		C. CONVEXITIES IN PROFILE	
Aire	1.09	North Scotland		Dart	-0.77
Calder (Y)	0.29	Dionard	1.31	Esk (Y)	-0.14
Dee	1.00	Hope	0.90	Fowey	-1.85
Derwent (Y)	2.54	Naver	1.41	Greta	-2.24
Don	0.47	Lake District		Teign	-0.11
Eden	0.21	Derwent	1.30		
Mersey	1.47	Duddon	0.51	D. DISCORDANCE	
Parrett	2.70	Irt	1.35	Ayr	-1.07
Severn	1.19	Kent	1.70	Ebbw	-1.33
Trent	1.00	North Wales		Elwy	-1.21
Witham	1.79	Dee	1.00	Exe	-1.05
		Dovey	0.88	Taff	-0.96
		Glaslyn	0.28	Tawe	-0.57

E. CHALK FLOORED		per cent chalk	
Avon (Dorset)	-0.74	(78)	
Colne (Herts)	-2.04	(75)	Where needed, the appropriate
Frome (Dorset)	-0.64	(55)	county is given, (Y) being
Kennet	-2.20	(80)	Yorkshire
Stour (Essex)	-0.26	(50)	
Test	-1.37	(75)	

Table 14.3 Stream profiles showing some departure from the degree of concavity typical of British rivers. The index is obtained from the formula for calculating standard errors, thus giving increasing weight to points at progressively greater distances from the general trend, However, since these departures are real rather than a matter of chance, probability statements are neither needed nor are they appropriate. Indices for 'over-concave' profiles are positive (A and B) whilst negative signs distinguish 'under-concave' profiles (C, D and E)

Infilling or aggradation may also be responsible for the over-concavity of another sub-group of profiles (Fig 14.3 and Table 14.3B). Mackin (1948) specifies rivers which were invaded by glaciers as presenting striking examples of aggradation and in drawing attention to the relevance of bed material, suggested that in such aggrading streams the downstream decrease in bed-load calibre was more rapid than in a graded stream, and, in consequence, such aggrading streams have profiles which are strongly upwards-concave.

Shape of longitudinal profiles

The rivers in Table 14.3B are readily allocated to one of three regions, the Lake District, North Wales or the North-West Highlands of Scotland, and there is ample testimony that all three were centres of ice dispersal (Clayton 1974). Phemister (1960), for example, has indicated that valley glaciation of the Dionard, Naver and Hope took place towards the end of the Devensian as the Scandinavian ice-cap withdrew northwards. North Wales and the Lake District have long been acknowledged as having supported independent ice-caps. The marked over-concavity of profiles from these areas is possibly attributable to the aggradation caused indirectly by the ice. Some degree of estuary infilling may also have occurred, and in these comparatively short streams the relative effect on profile form may be substantial. There also exists the possibility of glacial over-deepening by valley glaciers and George (1961), for example, has deduced that in North Wales most of the former glaciers were "... sufficiently powerful to over-deepen their containing valleys." This form of over-concave profile may also be distinguished from the first sub-group by the nature of the geology of the basins which in this case is almost exclusively lower Palaeozoic.

Apart from infilling, river diversion may be a significant factor associated with 'over-concave' profiles as this grouping includes two rivers which have experienced strikingly similar recent histories. The diversion of the Severn "... ranks with that of the Derwent as one of the major glacial modifications of the British drainage system ..." (Wooldridge & Morgan 1959). Both C. Lapworth (1898) and F. W. Harmer (1907) advanced the idea that the upper Severn and Vyrnwy had drained towards the Dee before glacial times. The diversion towards the Bristol Channel was attributed to the effect of a former ice-front which prevented the river's use of its pre-glacial lower course (Wills 1924). The retreating Devensian ice of the Cheshire Plain caused the impounding of the waters of the upper Severn, forming "Lake Lapworth" over the area now in the vicinity of Shrewsbury. These waters apparently escaped through what is today Ironbridge Gorge (Poole & Whitman 1961). Similarly the Yorkshire Derwent's course may have undergone recent and major re-alignment attributable to the Scandinavian North Sea ice blocking a former exit to the sea near Scarborough (Reed 1901). The present course to the North Sea, making use of the Humber estuary, is a much longer route. Reed argued that waters of the Yorkshire Derwent, unable to maintain their former short route to the sea, flowed southwards into the Vale of Pickering lowlands, forming "Lake Pickering" which, in turn, spilled over the low ridge of land at the eastern end of the Howardian Hills and created Kirkham Gorge.

Thus, from noting the similarity between river long profiles (Fig 14.4), the arguments to explain the interrelationship between "Lake Lapworth" and the Severn's Ironbridge Gorge and between "Lake Pickering" and the Yorkshire Derwent's Kirkham Gorge can be seen to be analogous. In addition, combined with the geological similarities of the lower courses of these two rivers over drift-covered

Permo-Triassic lowlands, another factor may be observed. Both the Severn and the Derwent, as a direct result of their diversions, have had their courses lengthened. This, by itself, could have increased profile concavity through simple geometrical modification. It is difficult to assess the extent to which the present profile forms are an expression of either infilling or of course extension. The Derwent appears to be about 4.2 times the length of its pre-glacial precursor. As the Severn's course is only 1.8 times as long, its degree of over-concavity is much less pronounced than in the case of the Derwent (Table 14.3A).

2. 'Under-concave' profiles

Of the profiles which are conspicuously less concave than the average profile form for British rivers, the extreme case of the River Greta (Fig 14.5) attracts immediate attention. The Greta rises in north-west Yorkshire and flows west-south-west for 26 km before joining the River Ribble, having fallen some 420 m within that distance. Examination of the profile shows that 'under-concavity' in the case of the Greta is due to the geometrical effect of the presence of a major convexity within its overall course. In actuality, the profile is composite with distinctly separate reaches developing above and below Ingleton. The convexity is linked with a local base-level close to the zone where the Craven Fault system crosses the valley and can be attributed to two main geological circumstances. First, there have been recurrent movements along the Craven Faults since Carboniferous times, with the 'rigid block' to the north of the faults having a positive tendency (Dunham et al 1953). Later movements were as recent as the Miocene (Bennison & Wright 1976) and, on the fault line itself, Giggleswick Scar has a freshness of form which would support an hypothesis of recent movement, and earth tremors in 1944 have had their epicentres traced to the fault lines. Jennings (1972) also advocated posthumous faulting in accounting for similar irregularities in the profiles of some of the rivers in the area around Canberra. Secondly, where crossed by the present channel of the Greta, faulting has thrown the resistant folded Ingletonian slates up against the much less resistant shales and sandstones to the south-west, downthrow side. This abrupt lithological contrast accentuates the effect of the structural dislocation on the profile.

Since the extreme case of the Greta indicates that an overall 'under-concavity' is attributable to a major local convexity within the profile, it is important to consider whether other profiles with major convexities also exhibit degrees of under-concavity as a result. As an index of profile convexity magnitude, the longitudinal sectional area of the convexity as revealed on the profile plots was used. An absolute scale for magnitude was preferred in this instance to focus initial attention on the physically large convexities. Of the 20 profiles with convexities larger than that of the Greta, 14 were under-concave (Fig 14.3), a proportion sufficient to suggest that

Shape of longitudinal profiles

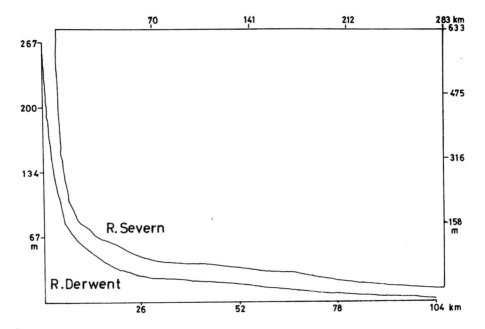

Figure 14.4   'Over-concavity' compared with the shape typical of rivers in Britain, illustrated by the River Severn and the Yorkshire Derwent.

the presence of a convexity may indeed influence the extent and rate at which hydrological processes can produce even an average degree of profile concavity. Of the six profiles with convexities larger than that of the Greta's but which were not 'under-concave' overall, five had convexities of a very broad span.

Of the 'under-concave' profiles which also incorporate an appreciable degree of convexity, a grouping of profiles in south-west England is noteworthy. The rivers Dart, Fowey and Teign are included in this group (Table 14.3C & Fig 14.3) and in each case the convexity is associated with the Permian granite bosses of the area. The Dart and Teign both drain Dartmoor and their gradients steepen where they leave the granite outcrop, as does the Fowey (Fig 14.6) where it leaves Bodmin Moor. Although less pronounced, convexities are also present in other river profiles in the south-west such as the Camel, Fal, and, to a lesser extent, the Taw. In each case the convexity is associated with the same change in lithology. The different altitudes at which these convexities are encountered suggests that they are due to such geological characteristics of the area rather than to some episode of geomorphological history. Strikingly, convexities are absent from the profiles of the Tamar and Torridge, neither of which come into direct contact with the granite. Possibly the granite, when once exposed to subaerial denudation, is less resistant to weathering than the rim of Devonian sandstones in the

Shape of longitudinal profiles

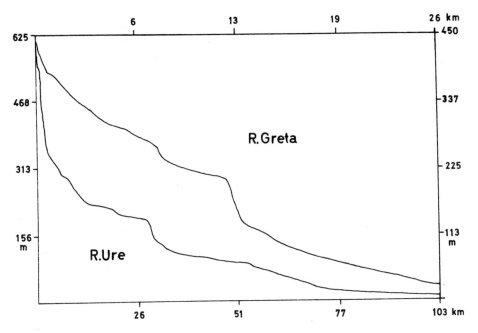

Figure 14.5 'Under-concavity' compared with the shape typical of rivers in Britain due largely to a 'local convexity' in the profile. In the River Greta, God's Bridge is 8 km from the source, the view in Fig 4.1 at 8.5-9 km, and the Craven Faults at 13 km. On the River Ure in Wensleydale, the Aysgarth Falls begin at 28 km from the source.

south and the Carboniferous sandstones and shales to the north wihin the metamorphic aureole surrounding the granites. The work of West (1978) has also considered the case of the Dart's profile and she provides a more detailed and alternative, though not necessarily contradictory, explanation for its characteristic form.

In other cases where a local convexity introduces a notable element of under-concavity into the overall profile shape, some striking change in the river's geomorphological history seems to be the significant factor, rather than lithological changes in the river's bedrock. For instance the overall 'under-concavity' of the River Ure is attributable in part to the marked convexity at Aysgarth (Fig 14.5) and the fact that a composite profile, with partially independent curves developing above and below Aysgarth, has been attributed to unusual changes in the river's geomorphological past. The present-day falls at Aysgarth could be due to a minor diversion at that point, possibly during the last advance of ice (King 1935). However, the dichotomy between the two component shapes which join at Aysgarth is so distinct that a geomorphological history of events requiring such longer spans of time than merely the post-glacial episode is suggested. Some form of stream long

Shape of longitudinal profiles

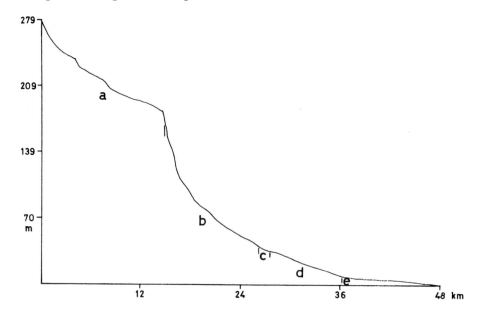

Figure 14.6 Longitudinal profile and stream-bed geology of the River Fowey. The stream-bed geology is: a) Granite (Permian); b) Slates (Middle Devonian); c) Staddon Slates; d) Meadfoot Calcareous Slates; e) Dartmouth Silts and Slates (all Lower Devonian)

profile discontinuity may well have existed at this point before the glacial periods, but this would have been on an old course of the Ure about 1.5 km to the north of its present channel (King 1935). The Yorkshire Esk (Fig 14.7) is another 'under-concave' stream which incorporates a relatively large convexity and a striking long profile dichotomy. The upper section is a smooth, upwards-concave curve, whereas the lower section is rugged and irregular by comparison, but disagreement exists concerning the precise history of the Esk valley during the Pleistocene period. Whilst Kendall (1902, 1903) argued that the western end of Eskdale was submerged beneath an ice-marginal lake, more recently K. J. Gregory (1961) has proposed that this western area was subject to stagnant ice conditions. Both agree, however, that certain major differences exist between the western and eastern ends of the valley and divide the valley at that point where the longitudinal profile changes character. Kendall (1902) considered that the Lealholm area marked the boundary between a glaciated and an unglaciated area. He thus interprets the Lealholm gorge as a feature caused by the diversion of the Esk due to the presence of a ice lobe in Eastern Eskdale. Gregory (1961), noting that east of Lealholm the Esk follows a course through several gorges possibly introduced as a result of the last glaciation, interpreted the channel gorge at Lealholm as a sub-glacial feature.

A further sub-category of 'under-concave' profiles may also be identified. The streams concerned (Table 14.3D) may be generally described as strongly discordant with the underlying structure. Kidson (1962), for example, has already noted the Exe's "... complete disregard for structure." He assumed that the Exe has yet to adjust itself to the underlying geology, which is extensively folded Carboniferous and Permian sandstone, as is seen in the deep incisions which the middle course is making into its bedrock. George (1970) has shown how discordant the streams of South Wales are in relation to the geology of their basins. The streams cut across a variety of rocks and structures almost perpendicularly. The South Wales coalfield is dominated by these periclinal flexures of the Hercynian syncline which runs from St. Brides Bay to Monmouth, but within this syncline a number of other, parallel, structures may be identified. The Ebbw, for instance, crosses both the Gelligaer and Llantwit-Caerphilly synclines and the Pontypridd and Usk anticlines. The Taff's course encounters the same structures with the exception of the Cardiff-Cowbridge anticline, which has replaced the Usk anticline as the dominant structure in the Old Red Sandstone. The Tawe's profile may be analogous, but the more complex recent history of drainage re-organisation in Neath and Towy basins may have obscured the relationship there. Similar arguments, but applied to extensive faulting rather than flexuring, may apply to the profiles of the Elwy (North Wales) and the Ayr. Embleton (1960) has already drawn attention to the complex faulting in the Silurian rocks of the Elwy's basin. Similarly, the Ayr's course, although crossing the resistant Permian Basalts, also encounters a large number of faults in an area of complex geology south-west of Glasgow (George 1958). In all of these cases the degree of profile 'under-concavity' is appreciable.

3. 'Under-concave' profiles with relatively constant gradients

The 'under-concave' profiles considered so far are of this overall form largely because they incorporate a major convexity attributable either to differing rock resistance or to events in geomorphological history. In contrast to this type of form, a distinctive group of profiles can be recognised which are less concave than the average British river simply because their profiles overall are relatively constant in gradient. Noteworthy examples, within the middle Thames catchment, at about the same altitude and of comparable drainage area, are the Hertfordshire Colne and the River Kennett (Fig 14.7). The facts that both these river basins are largely chalk-floored, 75 per cent for the Colne and 80 per cent for the Kennet, and lie within an area of rainfall appreciably less than the British average, indicate that some explanation for the distinctive profile shape might be linked with meteorological and hydro-geological conditions. Therefore regression equations were calculated to establish the rate of increase in runoff with increased basin elevation and increased rainfall in order to

Table 14.4 Calculations demonstrating the degree to which rainfall and runoff within the Thames Basin are, per unit area, below values estimated as typical for areas in the British Isles of comparable elevation.

A. RELATIONSHIPS

   expression

   rainfall = 85.17 (max. elev.)$^{0.413}$ $\underline{r}$ = 0.75

   runoff   = 2.031 (max. elev.)$^{0.918}$ $\underline{r}$ = 0.75

   runoff   = 0.0013 (rainfall)$^{1.93}$ $\underline{r}$ = 0.88

B. RESULTS FOR THE THAMES BASIN

independent variable	dependent variable	estimated value	actual value
max. elev.	rainfall	934	721
max. elev.	runoff	417	233
rainfall	runoff	411	233

maximum elevation (max. elev.), m; mean annual rainfall and runoff, mm.

examine the degree to which the Thames Basin rivers are hydrologically atypical (Table 14.4). These results indicate the degree to which the well-known 'rain shadow' of the Home Counties is reflected in below-average rainfall. Even taking the highest point in the basin as an expression of elevation, these rivers receive less precipitation than basins of comparable altitudes elsewhere in the British Isles. In addition to the relatively low precipitation, the low runoff probably also reflects the permeability of the chalk and the associated predominance of groundwater discharge.

It has already been suggested that the degree of profile concavity in British streams is associated with the rate of increase in discharge, higher values being linked with profiles with greater stream relief. It is consistent with this hypothesis that the converse should not only be recognisable, but also readily explained since the 'under-concave' profiles with relatively even gradients are developed in areas of permeable strata with below-average effective precipitation. Having identified this condition by examining the pronounced 'under-concavity' of the Colne and Kennet, it is noteworthy that four streams in largely chalk-floored basins have rainfall receipt and runoff values which are below values estimated as typical for the British Isles. In each case, the profile is 'under-concave' (Table 14.3E). In establishing the degree to which precipitation and runoff conditions are atypical in the chalk basins of south-east England, indices inevitably have to be based on the data for contemporary hydrometeorological conditions. Obviously, the use of this data is merely as an

Shape of longitudinal profiles

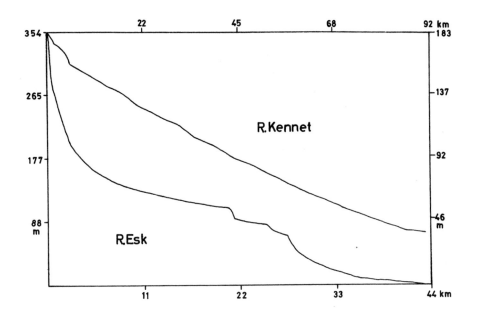

Figure 14.7 The 'under-concave' longitudinal stream profiles of the chalkland River Kennet and the Yorkshire Esk.

index of the cumulative, long-term conditions rather than an implication that the present-day profiles are 'nicely adjusted' to contemporary processes.

One further 'under-concave' profile in which the overall form is one of a relatively constant gradient is noteworthy. This is the Ramsdale Beck, which drains to the North Yorkshire coast at Robin Hood's Bay. Along this coast cliffs of Lias shales and sandstones continue to recede at a rate of 8.5-18.9 m/100 years (Agar 1960). As the Ramsdale Beck is only 5.4 km long, such rates of coastal recession would have appreciable impact on stream profile length, even within post-glacial times. By continuing to shorten stream length, the effect of coastal recession on Ramsdale Beck is comparable with a fall in base-level with continual steepening of the profile being an adjustment to these changing conditions. This example draws attention to the possibility that, under certain circumstances, geomorphological processes as well as permeable lithologies or relative low precipitation can produce relatively straight long profile shapes.

CONCLUSIONS

The main conclusion of the present enquiry is that, on the scales of streams developed in the British Isles, the degree of stream long profile concavity is systematically related to hydrometeorological conditions. That a meteorological gradient between west and east Britain can be sufficient in amount and persistence during geological spans of time to be reflected in morphological contrasts has already been described for glacial erosion features in the Scottish Highlands (Linton 1959). Counteracting this general tendency are several more particular influences which can be detected by tracing instances, which are points widely scattered from the regression line, back to the local conditions of their environments. Geological controls and changes in geomorphological history have both been shown to play a role, as have hydrometeorological settings. Throughout the discussion, a genetic terminology has been avoided. By using terms descriptive only of the observed forms of real shape, the ambiguities introduced by the genetic terms like 'rejuvenation' and 'knick points' or 'grade' have been avoided. As a result, the longitudinal profile of streams emerges as more than the cumulative effect of fluvial processes alone, being a theme and thread which runs through glacial, fluvio-glacial and coastal processes too.

REFERENCES

Agar, R., 1960. Post-glacial erosion in the North Yorkshire coast from the Tees estuary to Ravenscar. *Proceedings of the Yorkshire Geological Society,* 32, 409-428

Bennison, G.M. & Wright, A.E., 1976. *The geological history of the British Isles.* (Arnold, London)

Boswell, K.C., 1946. A detailed profile of the River Test. *Proceedings of the Geologists' Association,* 57, 102-116

Brown, E.H., 1952. The River Ystwyth, Cardiganshire: a geomorphological analysis. *Proceedings of the Geologists' Association,* 63, 244-269

Clayton, K.M., 1974. Zones of glacial erosion. In: *Progress in geomorphology,* ed. E. H. Brown & R. S. Waters, Institute of British Geographers Special Publication 7, 163-176

Dunham, K.C., Hemingway, J.H., Versey, H.C. & Wilcockson, W.H., 1953. A guide to the geology of the district around Ingleborough. *Proceedings of the Yorkshire Geological Society,* 29, 77-115

Embleton, C., 1960. The Elwy river system: Denbighshire. *Geographical Journal,* 126, 318-334

George, T.N., 1958. Geology and geomorphology of the Glasgow district. In: *The Glasgow region,* ed. R. Miller & J. Tivy, British Association for the Advancement of Science, Glasgow, 17-61

George, T.N., 1961. *British Regional Geology: North Wales,* (Her Majesty's Stationery Office, London), 3rd edition

George, T.N., 1970. *British Regional Geology: South Wales* (Her Majesty's Stationery Office, London), 3rd edition

Godwin, H., 1960. Prehistoric wooden trackways of the Somerset Levels. *Proceedings of the Prehistory Society,* 26, 1-36

Gregory, K.J., 1961. Proglacial Lake Eskdale after 60 years. *Transactions of the Institute of British Geographers,* 36, 149-161

Gresswell, R.K., 1964. The origin of the Mersey and Dee estuaries. *Geological Journal,* 4, 77-86

Harmer, F.W., 1907. On the origin of certain canyon-like valleys associated with lake-like areas of depressions. *Quarterly Journal of the Geological Society,* 63, 470-514

Hollingworth, S.E., 1931. The glaciation of western Edenside. *Quarterly Journal of the Geological Society,* 87, 281-359

Jennings, J.N., 1972. The age of Canberra landforms. *Journal of the Geological Society of Australia,* 19, 371-378

Jones, O.T., 1924. The longitudinal profiles of the Upper Towy drainage system. *Quarterly Journal of the Geological Society,* 80, 568-609

Kendall, P.F., 1902. A system of glacier lakes in the Cleveland Hills. *Quarterly Journal of the Geological Society,* 58, 471-571

Kendall, P.F., 1903. The glacier lakes of Cleveland. *Proceedings of the Yorkshire Geological Society,* 15, 1-40

Kidson, C., 1962. The denudation chronology of the River Exe. *Transactions of the Institute of British Geographers,* 29, 43-66

King, W.B.R., 1935. The upper Wensleydale river system. *Proceedings of the Yorkshire Geological Society,* 23, 10-24

Langbein, W.B., 1964. Profiles of rivers with uniform discharge. *United States Geological Survey Professional Paper,* 501-B, 119-122

Lapworth, C. et al., 1898. Long excursion to the Birmingham district. *Proceedings of the Geologists' Association,* 15, 417-428

Leopold, L.B. & Langbein, W.B., 1962. The concept of entropy in landscape evolution. *United States Geological Survey Professional Paper,* 500-A

Lewis, W.V., 1945. Nick points and the curve of water erosion. *Geological Magazine,* 82, 256-266

Linton, D.L., 1959. Morphological contrasts of eastern and western Scotland. In: *Geographical essays in memory of A. G. Ogilvie,* ed. R. Miller & J. W. Watson (Nelson, London), 16-45

Mackin, J.H., 1948. Concept of the graded river. *Bulletin of the Geological Society America,* 59, 463-512

Phemister, J., 1960. *British Regional Geology: the Northern Highlands,* (Her Majesty's Stationery Office, London), 3rd edition

Poole, E.G. & Whiteman, A.J., 1961. The glacial drifts of the southern part of the Shropshire-Cheshire plain. *Quarterly Journal of the Geological Society,* 117, 91-130

Reed, F.R.C., 1901. *The geological history of the rivers of east Yorkshire.* (Clay & Sons, London)

Rubey, W.W., 1933. Equilibrium conditions in debris-laden streams. *Transactions of the American Geophysical Union,* 14, 497-505

Shulits, S., 1941. Rational equation of the river bed profile. *Transactions of the American Geophysical Union,* 22, 622-630

West, E., 1978. *The equilibrium of natural streams.* (Geo Books, Norwich)

Williams, J.C., 1968. The buried channel and superficial deposits of the Lower Usk. *Proceedings of the Geologists' Association,* 79, 325-348

Wills, L.J., 1924. The development of the Severn Valley in the neighbourhood of Ironbridge and Bridgenorth. *Quarterly Journal of the Geological Society,* 80, 274-314

Wills, L.J., 1938. The Pleistocene development of the Severn. *Quarterly Journal of the Geological Society,* 94, 161-242

Wooldridge, S.W. & Morgan, R.S., 1959. *An outline of geomorphology: the physical basis of geography.* (Longmans, London)

CHAPTER 15   CONCLUSIONS

Alistair Pitty

GEOMORPHOLOGICAL IMPLICATIONS

1. Traditional roots

The history of geomorphological studies in the British Isles is made up of four interwoven strands. The first three are considered presently, whilst the fourth - the interdependence of geomorphology and geography - is taken up later. A central strand of geomorphology in the British Isles, if not its backbone, stretches back through time to the 19th century geologists, like A. C. Ramsay, A. J. Jukes-Browne and J.Playfair, who offered interpretations of the land forms they saw around them. Episodically, in part the product of a common language, external influences have been largely those of American geologists, as the writings of A. N. Strahler or L. B. Leopold continue to demonstrate. Thirdly, and in contrast to the work of most American geologists contributing to landform studies, the range of environments in which geomorphologists from the British Isles can or have worked abroad introduces a particular geographical width and latitudinal range.

All these three strands are evident in the present book. The link straight back to the observations and descriptions of the 19th century geologists like A. Geikie is the more immediate because our efforts, with their emphasis on actual observations, are unaffected by the lengthy diversions of the deductive, model-building mentality introduced and embedded by W. M. Davis. The stimulating influence of other American geologists, however, is quickly perceived by inspection of the bibliographies of chapters in this book. Finally, the geographical span of our enquiries is also self-evident in the distinctiveness of the chapters based on original research in South-east Asia, Australia, Africa, and continental Europe.

2. Geological variability

Despite traditional roots and training in geology, many geomorphological enquiries postulate homogeneity of lithology and idealise other geological features. In contrast, the present approaches, with their emphasis on the actualities of the environment, inevitably repeatedly reveal the importance of geological heterogeneity at all scales. The differences between the solutional processes of siliceous rocks and calcareous rocks are exemplified. The significance of stratigraphy in juxtaposing contrasted lithologies is stressed in the studies of chemical

Conclusions

weathering in North-west Yorkshire and in the demonstrable susceptibility of the land surface to mass-movement in the North York Moors. Lithology also influences the regolith characteristics on Dartmoor, beneath the Luxembourg forests, the mineralogical attributes of alluvium in Kenyan stream channels and the varying proportions of pebble type in the Upper Wharfe. In the case of stream longitudinal profiles, not only is contrasted lithology and varying dispositions of rocks repeatedly in evidence, but so also are changes in the geological past, including structural movements. Beneath the ground, the role of joints, as the collecting ground for the actual 'headwaters' of many streams, is inferred for the Dartmoor granite and is demonstrated in the limestones and other jointed rocks in northern England.

3. The biological factor

In keeping with the approach of American geologists to fluvial geomorphology, the biological factor has not been identified under a major heading. Nonetheless, our detailed investigations of fluvial processes in a semi-natural context reveal the degree to which biological phenomena are an integral part of chemical and physical fluvial processes. The role of vegetation cover in fluvial processes before water even reaches the ground is clearly established in the studies on Dartmoor, in Luxembourg and in the North Queensland rainforest. The vegetation cover is also seen to be an integral part of the processes by which silica is released in solution from areas of granite weathering. The importance of litter cover is established for both chemical weathering of Malaysian limestone outcrops and for the physical dislodgement and entrainment of soil particles on the Luxembourg forest floors. It follows that where the organic matter accumulates to the extent that peat is formed, the initial phases of runoff are greatly modified compared with those observed on the bare ground of semi-arid environments.

The significance of the animals which toil directly at and below the ground surface is also observed. Biopores are critical in the subsoil drainage at all levels in the experimental pits excavated on Dartmoor. On the forest floors in Queensland and in Luxembourg, the major role of animals in disturbing soil is described. Not least the humble role of snails, as an integral part of solutional processes on bare limestone outcrops is also illustrated. In keeping with the increasing range of technological accomplishment, the role of man as an integral part of fluvial processes is specified, particularly in accounting for the varying erosion rates observed in the catchments in the North York Moors. The significance of vegetation changes introduced on Dartmoor are also examined in detail and the effect of artificial drainage investigated in one of the Luxembourg catchments. The susceptibility of peatland to change with man's incursions is even highlighted by the experimental design of investigations in the Brishie Bog, since man's weight alone can modify water levels in peat.

4. Scale

In an exercise involving much measurement of actual ground-surface phenomena, the significance of geographical scale is particularly clearly demonstrated. Although quantities of weathered materials delivered to an ocean depend on the drainage area of streams, the relative importance of the several factors controlling fluvial processes changes according to scale. The point is examined specifically in the case of the factors influencing denudation in the North York Moors. Ultimately, the morphological inter-relationship in land forms produced by fluvial processes in channel pattern or in longitudinal profile shape is seen to be related to scale. It is seen that the generalisations of American geologists, based on the characteristics of phenomena on continental scales require careful testing against the scale of land forms developed within a relative small island. Scale is also a significant influence on methods of enquiry. Due to the correlation between depth and width there is a literal upper limit to the size of meander bends which can actually be measured by an individual worker. In contrast, the scale of longitudinal profiles makes the field survey of only one or even part of one stream profile feasible. Therefore, a study of significantly large number of individuals depends inevitably on map-based data and any insistence that geomorphological data must invariably be based on field data is necessarily over-ruled. Conversely, scale can be used as a technique with which to simplify deliberately but without artificially the complexity of environmental relationships, as the study of solutional weathering on rock outcrops in Malaysia demonstrates.

THE PRESENT APPROACHES AND DEFINITIONS OF GEOGRAPHY

As each of the approaches used here has proved fruitful, the degree to which the adjective 'geographical' is appropriate should be evaluated. As there are many differing emphases in statements defining 'geographical' studies, such comparisons are not straightforward. Indeed, if there were a widely accepted, rigorous definition of geography, such an exercise would be superfluous. The present context is, however, particularly propitious for making such comparisons since most general pronouncements on geography are not linked with worked solutions to actual investigations. Too readily, the would-be programmers of geography overlook the fact that the viability of any approach or method is only proven by its workability.

1. Richard Hartshorne's view of Geography

A conspicuous landmark in the methodology of Geography for English-speaking students is the work of Hartshorne, with the value of his *The Nature of Geography* (1939) being re-inforced by the appearance of his *Perspectives on the nature of Geography* 20 years later. A synthesis of these works could be a particularly reliable rule against which to measure the value of methodological statements, for

Conclusions

several reasons. Obviously, the time span is valuable, with the *Perspectives* allowing opportunities for re-affirmation, change, or even reversal of earlier conclusions. The time span covered is, in effect, much longer than 20 years. Hartshorne depended heavily on the writings of Alfred Hettner who, in turn, had set himself the task of providing a sound methodological exposition of Geography as it had evolved in Germany in the 19th century. By comparison some writings which have followed Hartshorne's *Perspectives* could prove to be comparatively transient, with impulsive advocates for 'change' merely strengthening the resolve of those preferring to resist change. This possibility is strengthened by the case presented by Christopher Booker (1969) in *The Neophiliacs* in which he argues that adulation of the 'new' was a peculiarity of the everyday culture of the 1960's.

To compare the paths traced in the present book with the route clearly signposted by Hartshorne has a further, particular advantage. Even the briefest observations on geography tend to propound definitions which, from the outset, appear simply to strengthen the significance of the specialist position of the authors concerned. Hartshorne's statements were much dependent on German sources but written in North America by a human geographer with particular interests in a regional approach to political and to agricultural geography. In contrast, our systematic studies of the natural environment depend heavily on the writings of American geologists. Furthermore, they were written in Europe and Australia, and by earth and environmental scientists as well as by physical geographers. In view of these contrasts, sections where our paths follow or even parallel Hartshorne's route could be particularly significant, whilst the opportunities to bend geography merely to support our predilections are minimized.

2. Landscape as an object of study

According to Hartshorne (1939, page 262) "geography has at least one individual unitary, concrete object of study, namely the whole earth". However, it has always been recognised that geographers, concerned to identify a specific object for their studies, have found that all possible objects are more adequately accounted for by the definitions of more specialised fields, if not already self-evident in their titles. In this context 'landscape' has been a term much used by geographers. Derived from the German geographers' *Landschaft*, the concept was expressed initially in the writings of Passarge, Brunhes and Sauer. Today geomorphologists cannot ignore the claims that the 'landscape' is a proper object for study if Leopold, Wolman & Miller (1964) also observe that "... in the landscape, the prime sculpturing agent is running water". Indeed, many points of similarity between our experiences and the definitions of those whom Dickinson (1939) described as the 'landscape purists' can be identified. In seeking to limit geography to objects that are perceptible to the senses, whether those of sight, sound, smell or feeling, the 'landscape purists' identify exactly our field experiences.

Conclusions

The sight of the landscapes studied is evident in our photographs. Equally sensed, however, is the sound of wind in trees and grass, the rush of turbulent water or droplets hitting cave or forest floor, the smell of damp or drying sand and soil or decaying organic matter, the dankness of forest, fissure, and soil pit. Finally there is the feel of frozen soil, hot sand, the constriction of fissure or pit, the push and drag of streamflow. Not least there is the downpour, with the authors of Chapter 5 laying jocular claim to belong to the only forest hydrology research team to wear life jackets whilst at work.

There are fundamental contrasts. The landscape purist, in seeking to define the study of landscape as specifically that of the geographer, emphasized material objects, a focus on form rather than on process. Thus Hartshorne initially concluded that the majority of facts which geographers can observe directly are by nature static. This is one of the few views which he later reversed in the *Perspectives* in which he states that the geographer's concern is " ... not merely with static earth features but with those in motion, whether air currents, streams, or the human activities of transport which are the essential factors in integration among places". First, as in any study concerned with processes, the insensible is clearly an important if not a predominant consideration in our studies. The second contrast is largely visible, less obvious, but equally fundamental. In identifying the appearance of the landscape as the object of their study, the 'landscape purists' referred to the outermost visible surfaces, whether this was the tree tops, the walls of buildings or water surfaces. The soil was ephemeral, depending on the seasonality of crop growth. In contrast, our considerations range from depths of groundwater circulation to the upper atmosphere. In striking contrast, our level of interest is consistently below the physical surface of the vegetation, soil, or water. In some cases it is even below the uppermost surface of bedrock.

If the portion of our overlap with the landscape purists means that, similarly, no objects distinctively for the geographer's study are identified by their concepts, the sensations of landscape are nonetheless critically significant to our approaches. Intensive or recurrent study in the same area leaves not merely geomorphological impressions. Images of the landscape, the sky by day and night, the people who come and go, the memories of those who came and went, all form a unified whole in the individual mind. The impression that one has merely studied intensively part of a whole is indelible. The result of this 'experience of landscape' (Appleton 1975) is that neither the regional-systematic nor the physical-human forms of dualism emerge. Dualism is a perennial source of anxiety only in the minds of geographers abstracted from reality.

3. <u>Man and Nature</u>

The considerations of landscape in geography has enlarged to include the study of the origin and evolution

Conclusions

of landscapes. In contrast, other geographers who conclude that the search for a unique *object* for disciplined study was fruitless, have emphasized the study of relationships between earthly phenomena instead. Of the many possible pairings in such relationships, those between Man and Nature have been most widely claimed as germane to a distinctive definition of geography. To avoid tracing either 'landscape' or 'Man-Nature' themes backwards indefinitely into the geological past, a starting point where 'man entered the scene' is needed. A range of scientific and historical techniques yields only vague suggestions compared with the beliefs of followers of the religious teleological viewpoint of Ritter. It is taxing enough for geographers embracing the Man-Nature theme to stand astride the man-made cleft between the natural and the social sciences. It is an added unease that, at the base of this cleft, an undercurrent of views on the origin of man flows from both teleological and evolutionary sources, to mingle like oil and water. Strong words, however, have helped to bind the Man-Nature theme securely together. For example, " ... the separation of the whole into man and his environment is such a murderous act ... It is like studying a human being without his nervous system" (Herbertson 1916, page 149). Preston James (1934, page 79) was one of those not anxious to swing, describing the contrast between Man and Nature as 'supreme arrogance'. The concept of 'causality' lends an additional binding strand for geographers perceiving some fundamental difference between Man and Nature. On the one hand there was the determinists' view of man dominated by mechanistic laws of nature. With technological advances, this viewpoint faded from geography, through increasingly pale shades of 'environmentalism' and was last clearly seen in the works collected by Antarctic geologist Griffith Taylor (1951). Even the 1957 edition of his *Geography in the twentieth century* continued to immortalise Ellsworth Huntington's embarrassingly non-Canutian words - "Enthusiasts say that the aeroplane has overcome these natural barriers, and that we shall have regular service by air, not only across mountains, but also across the practically uninhabited forests, tundras, mountains, glaciers, and ice floes of northern Canada, Greenland, Alaska, northern Asia and the Arctic Ocean. I do not share this idea. It seems to me to be out of harmony with the established principles of geography". This quotation illustrates why geographers were particularly keen to disinvest themselves of such viewpoints. The irony 'on the beach' of geography in the 1970's results, however, from the environmentalists' tide running completely out, whereas, in the wake of a gale of technological excesses, it has surged back over so many adjacent strands. In geography, however, there remains " ... the inversion of the classical system, the view that geography is the study of the effect of Man on land, of the effect of the material culture of societies on local ecological, hydrological and other systems" (Wrigley 1965, page 13). Viewed in either direction, the 'causal relationship' linking theme has the fundamental weakness of partiality. That some 'relationship' exists is a basic premise, so that the direction in which any enquiry might find success or failure is predetermined from the start. Subjective also is the use of the word 'man' himself. Is 'man' a commuter, a Kalahari bushman, a Hindu, or one of a Peoples' Republic? There is also a near 50-50 chance that man is, in fact, a woman, and in this sense, Elaine Morgan (1976) offers opinions on urbanisation which are in striking contrast to the megalopolis of male views of the city.

It is awkward for geographers that set against strong words, preconceived notions about causality and the nature of man, and the administrative convenience which binds their subject together, that it is the procedure of the systematic sciences to study categories of phenomena as much as possible in isolation from each other. More generally, philosophers also find it helpful to distinguish that which is man and that which is not, if only to facilitate a clear exposition of intricate relationships.

Insofar as geographers have assumed that their subject is distinctively equated with the study of inter-relationships between Man and Nature, there is the temptation to overlook the fact that most natural sciences, quite unselfconsciously, have taken account of man in their studies. Looking at geography from the point of view of the philosophy of science, Kraft concluded that man and his works occupy a more prominent place in geography than would be expected from standard definitions of the subject (Hartshorne 1959, page 41). For a specific example, one need look no further than the classic statement on fluvial geomorphology by Leopold, Wolman & Miller (1964) published as a title in "A Series of Books in Geology" and in which the word 'geography' appears no more than half-a-dozen times. These authors observe " ... the obvious economic interest in rivers for drinking water, industrial use, irrigation, and navigation" (page 46). There are several references to engineering structures and problems (eg. pages 147, 298, 385). They recall how in the early days of boating on the Mississippi River, a large log was commonly attached to houseboats. Such water-soaked logs were commonly submerged and kept to a downstream course near the centre of the river (page 282). On one side of the Man and Nature coin, they explain that " ... owing to the flat slopes characteristic of the immense alluvial valleys of the Indian subcontinent, canal designers have always been forced to design within the constraint imposed by the natural valley gradients, which are very flat" (page 272). On the other hand, they observe that " ... there are cases where the degree of human influence is over-riding" (page 434), that " ... where severe overgrazing or noxious gases from industrial plants have denuded hillslopes of vegetation, accelerated erosion has created new channels" (page 427). Throughout there is an affectionate regard for mortal man, explicit in the statement that " ... the changes in regimen associated with construction of dams are certainly heroic" (page 457). Man the Geographer, however, is not identified, for "To both the riverman and the geologist, the depositional character of sand deposited under various

Conclusions

conditions of bed configuration is of importance" (page 226). For these authors it is neither necessary to use the word geography nor to insert the word 'man' into every paragraph, to demonstrate that mankind is an integral part of their viewpoint.

Despite the many pitfalls on the Man and Nature trail, it is hoped that in the present collection, people are free to come and go as they please, that man's environmental deeds and needs are neither blindly ignored nor arbitrarily emphasised. Although suppositions about causality are largely absent, advantage is taken of planned land-use in selecting study areas, as in the contrasted vegetation covers in the Dartmoor example (Chapter 2), the retention of some forest in Luxembourg (Chapter 6) and the restricted access to the peatland Nature Reserve (Chapter 9). Where analysis reveals that man's activities are a significant factor in explaining certain characteristics, this is indicated, as in the case of borehole data (Chapter 4) or in the multiple regression equations of sediment loss (Chapter 8). The fact that all environments studied are, in varying degrees, less than the purely natural of the Garden of Eden, is a problem only where a hiatus between Man and Nature is presupposed. In contrast, we are simply studying the actualities of present-day reality in areas beyond the urban fringe.

One problem remains. Within the field of geography many social scientists find it convenient to ignore studies of the natural environment unless 'relevance to man' is reiterated and spelt out in elementary, non-technical terms. In this situation it is not enough to agree with Leopold, Wolman & Miller (1964, page 46) that such things are 'obvious'. When confronted with the affectation of intellectual blindness, it is not enough loudly to give a dogma a bad name. Perhaps, therefore, we should indicate that the unique solutional weathering patterns on limestone (Chapter 3) are currently a source of much social friction in north-west England between landscape gardeners and conservationists. Peat erosion (Chapter 9) has accelerated in recent years in the southern Pennines to reduce the capacity of the Longdendale reservoirs at an alarming rate. We could use the catastrophic for emphasis, with a Wharfedale flood of 1686 "casting up water to a prodigious height", with bedload smashing through the village of Starbotton and covering acres of land"(Pontefract & Hartley 1938, page 64). In the case of bog bursts (Chapter 9), the Killarny flow of 1896 occurred above the head of a small stream and on the inquest of the family killed when their home was swept away " ... evidence was given that a 'wet vein' existed in the bog continuing the direction of the stream" (Sollas et al 1897). In the case of the more permanent fluvial features of meander plan and profile shape, references to cultural, urban, and industrial heritage could be re-iterated. The town cores of places like Durham and Shrewsbury are circumscribed by meander loops. In southern England, many of the old watermills " ... were located on or near knickpoints, and many of the riverside towns such as Canterbury, Winchester

Conclusions

and Norwich are sited in significant relation to them" (Wooldridge & Kircaldy 1936, page 5). Perhaps reference to culture and appeals to civilization should be taken further back in time. In the area studied in Chapter 11, research led by Dr. Mary and Richard Leakey has demonstrated that the sedimentary exposures to the east of Lake Turkana contains a unique fossil and artefact record that reveals evidence of early man during the Pliocene/early Pleistocene and of his activities. One of the most important aspects of archaeological investigation in this locality has been " ... recognition of the significance of the relationship between the location of archaeological sites and their paleogeographic setting in understanding early Pleistocene man/land relationships" (Harris & Herbich 1978, page 530). Sediments were an integral part of this environment, since the boulders and cobbles deposited as conglomerate in the beds of river and stream courses were readily available as a source of raw material for artefacts and had particularly good flaking qualities. "The accessibility and availability of raw materials appears to have been another critical factor in the decision to locate sites close to stream courses" (Harris & Herbich 1978, page 540). Appeal could even be made to a combined sense of civilization and the shock of catastrophe, for it seems that a flash flood, as depicted in Figure 11.5, was the first known flood in which hominids perished. At the Afar Locality 333, a group of five to seven adults and children of the genus *Homo* appear to have been struck down, some 3 million years ago (Johanson, 1976). "Some natural catastrophe" was Dr. Maurice Taieb's view, "I think maybe a flash flood, perhaps while they were resting or sleeping".

No such remarks are offered, perhaps because the definition of geography as "the study of the earth in relation to man" appeals to a " ... curiously oblique motivation of the investigations of a great class of natural phenomena"(Leighly 1955, page 310). In Chapter 10, it is noted how springs feed village ponds and fish farms, so that readers can test their reactions to such remarks. The next step in this direction would be to quote *Mainsprings of Civilization* by Ellsworth Huntington (1945) which shows how far down the slippery convex slope of causality the preceding paragraph has ventured. To be strictly accurate, the present book can lay little claim to contribute to the Man and Nature theme. Every chapter, however, does to some extent fulfil John Leighly's wish that "It would be good if we could again approach the earth with unhampered curiosity and attempt to satisfy that curiosity by whatever means the problems we encounter suggest" (Leighly 1955, page 318).

4. Geography and Time

It is evident that time is an integral part of the Man and Nature theme. Indeed, for those geographers who see their study as being primarily concerned with the changes in the phenomena under investigation, chronology and temporal relations are a major guiding principle. The

Conclusions

fourth dimension is particularly marked in the definition and priorities of geographers whose early background training was in either history or geology, and where the context of time has been seen as one in which physical and human geography might be joined. Thus S. W. Wooldridge (1951, page 28) explained that "Geomorphology is the historical geography of the physical landscape" and that modern geography, being concerned with the physical landscape and with the cultural landscape, found that " ... its proper aim is to bring them into synoptic and stereoscopic review" (Wooldridge & Linton 1955, page 3).

In the present volume, it is phenomena as they exist at present which are the focal point of interest. The 'approaches' are explicitly geographical, not chronological, and therefore time is, by definition, the fourth dimension. Each phenomenon studied, however, is inevitably seen within varying spans of time. At the shortest span, the past is already the beginning of this sentence. At the longest span during which the more permanent features of fluvial processes have developed, current processes do not account for all the cumulative effects which explain the present morphology. This is particularly the case for long profiles (Chapter 14) but also plan forms shift and change in cycles of decades and centuries. A time span of days and months is needed to obtain a representative picture of sediment and solute characteristics and of soil water behaviour.

Within the time span during which present-day processes and forms are under observation, the sequences of chronological changes become, in themselves, a focal point of interest for many students of the natural environment. Geographers, however, " ... are forced to distinguish between an historical and a geographical point of view, and in order to master the technique of either, we need to keep clearly in mind the distinction between the two" (Hartshorne 1939, page 188). In the present instances, as repeatedly demonstrated on limestone pavement, forest floor, peatland, springs and flash floods, the focal point of interest is not in changes through time *per se,* but rather that fluctuations of temperature and rainfall in time provide a range of contrasting sets of natural inputs into the natural systems under observation. Only where a programmed geography has been deliberately welded together by time can the logic that time is a secondary consideration to those of the spatial dimensions be over-ruled.

5. The range of attitudes to heterogeneity at the earth's surface

Since the range of phenomena that might be considered 'geographical' is so vast and variable, some criteria are desirable for distinguishing those of greater significance from those of less. Thus serious students of geography must find some logical and consistent basis for identifying for study fewer than all variables present, and within areas of scales much less than that of the earth as a whole.

Conclusions

   a) Regional geography. A long-standing traditional view of heterogeneity was that the very nature of geography requires some basis of division of the earth's surface. The influential emphasis on progressive subdivision probably stems from Hettner, for whom the need in geography was " ... to comprehend the earth surface as a whole in its actual arrangement in continents, larger and smaller regions, and places" (Hartshorne 1959, page 13). Once a particular geographical area has been delimited, the purpose of geography becomes the comprehension of the integration of diverse phenomena and the varying ways in which they occur in areas. The present enquiries are consistent with this purpose in that we consider diverse phenomena in areas, but regardless of the conventionally defined fields of knowledge into which they might fall. The geographical areas studied, however, have been selected in a manner quite contrasted with that of progressive subdivision. The study areas have been selected with strict and deliberate attention to the distinctive characteristics of the chosen areas. Subsequently, tentative steps towards wider conclusions are made if results from studies in other areas afford some basis for comparison.

   b) Geographical distributions and spatial analysis. As evidence of a background in history or geology has faded from statements on the nature of geography, the historicists and chronologists have been replaced by the spatial analysts. Concurrently, the indefinability of regional boundaries has favoured increased attention to distributions of phenomena rather than the study of discrete areas. The view has gradually strengthened, therefore, that geographers, in contemplating the heterogeneity at the earth's surface, study the areal distribution of several different phenomena and their spatial inter-relationships. Indeed "Study of the spatial patterns displayed in the map is one of the essentials of geography" (Berry 1964) and "The integrating concepts and processes of the geographer relate to spatial arrangements and distributions, to spatial integration, to spatial interactions and organization, and to spatial processes" (Berry 1964, page 3). However, in noting that variation is temporal as well as spatial Berry then discusses time as *A Third Dimension*. Perhaps this misconception itself encourages a cautious view of applying approaches spatial to the natural environment, since it is man's distinctive trademark to organise space. From the Central Business District to the moorland wall, snow-line, or desert fringe, regularity of arrangement and functional relationships between widely separated places makes spatial analysis inevitably a man-centred if not an urbanocentric viewpoint. Nonetheless, there is a distinctive spatial sequence in fluvial geomorphology, following the gradients of water flow. This is seen in several of the present chapters, especially in the hydrological pathways defined in the Dartmoor study (Chapter 2). However, if spatial analysis is so tied to the plan view on maps of urban areas that time appears to be the only third dimension, spatial sequences in the natural environment, being unidirectional and linear relationships, are best displayed in cross-

Conclusions

section. This is the case for soil catenas and slope profiles (Chapter 7) and explicitly the case in the long profile (Chapter 14). Perhaps it is the insistence that spatial analysis is tied to a map which explains why so few geographers have offered studies of downstream sequences like those demonstrated in Chapter 11 and Chapter 12. Although the spatial characteristics revealed there, and in Chapters 2 and 14 also, are fundamentally different from those in urban areas, it is equally logical to describe such approaches as intrinsically 'geographical'.

c) *Exploration, discovery and travel.* The variability at the earth's surface has, in general, always beckoned curious travellers forward. More specifically, the remoteness, inaccessibility and the unknown of certain places has been defined as a particular challenge to man's capabilities. For some geographers, further perplexing dualisms are identified between the dilettante traveller and the scientific explorer, between the stupendous attainment of the inaccessible and the learned theorist in his library, globe to hand. For some academic geographers, like Richard Hartshorne, the exclusion of geographical explorers from their spheres is so emphatic that the issue is not even raised. Collectively the present contributions span sufficient of the middle ground between such extremes that the dichotomies between them could be briefly considered.

It is true that the glow of technical accomplishments by pioneer explorers was often quickly dimmed by events which followed in their wake, ranging from the unpleasant and impoverishment to the abominable. Geography, however, may be the poorer if deprived of the epic example of the manner in which early explorers put theories about the heterogeneity of the earth's surface to the test. James Cook, for instance, went to great lengths and accumulated innumerable detailed observations and measurements to limit the scope assumed by "speculative fabricators of geography". If names such as Cook's are thoughtfully bestowed on modern universities (Courtenay 1979), it seems that a geography which excludes exploration and discovery has omitted some vital values. On the other hand, in the mere two centuries since Cook's death, population numbers have greatly increased and spread, and the number of unvisited places has dwindled dramatically. For the price of an armchair, the traditional location for the inactive geographer, one can pass quickly over the places which repelled even flights of fancy in Huntington's mind only a generation ago. To search today for unvisited places is to mine a seam nearing exhaustion. If exploration and discovery are to remain vital cogs in the geographer's machine, a re-definition of the object of such endeavours requires priorities and values which can be recycled with little loss in worth. One ingredient in our process studies, with the emphasis on observing a range of natural inputs, is repeated travel to the same sampling spots. Indeed, if our total journeys were laid end to end, they would stretch far beyond the limits of the average taxpayer's patience. More seriously, with the importance of locating naturally simplified areas being a basic criterion in the formulation of the present approaches, many examples of challenges and rewards are offered to those who wish to scour the earth's surface. In this context, it is noteworthy that geologist Geikie's enthusiasm for the scientific opportunities for studies at the Falls of Clyde were not diminished by social engineer Robert Owen setting up his contrasting experiments at the same spot a couple of generations previously.

SCIENTIFIC, STATISTICAL, AND GEOGRAPHICAL METHOD

1. Baconian principles

Throughout the present book there are many examples of procedures adopted in which a conventional grasp of 'scientific method' is applied to studies of realities in the natural environment, despite the absence of laboratory controls with which scientific method is usually linked. The procedure that, with a given set of inter-related variables, all but one are held constant by artificial controls is an approach initially advocated by Francis Bacon, the Renaissance encyclopedist, lawyer, and courtier. "Let everything respecting natural bodies and virtues be, as far as possible, numbered, weighed measured, and defined", he urged (Anderson 1948, page 262). "Men who would invent useful things " ... ought to observe attentively, minutely, and systematically natural works and operations one by one ... " (Anderson 1948, page 287). He warned against conclusions where there had been " ... a failure to collect particular observations of sufficient number, variety, and certainty to inform and enlighten the understanding" (Anderson 1948, page 93). However, as soon as Bacon's exhortations were widely adopted in the explosion of scientific endeavour in the second half of the 17th century, scientific method became synonymous with laboratory control. Any procedure producing data to fewer decimal places was unquestionably less reliable and purely qualitative approaches counted themselves out altogether. This ranking of priorities has never favoured the serious measurement of the intrinsically heterogeneous phenomena in the natural environment, if observed in their natural setting with as little modification of the inter-related variables as possible. It may seem impertinent to postulate that perhaps what such studies may lack in terms of decimal places could be counterbalanced by the degrees to which they approach more precisely the actual nature of reality. However, they have as much claim to be described as 'Baconian' for a very simple reason. Laboratory experiments were a post-Baconian phenomenon. Simply because he lacked the facilities and the science which he advocated as necessary to alleviate the human condition, Bacon had inevitably to draw examples to illustrate his reasoning and to support his arguments from observations of the natural environment.

It may seem fatuously presumptious to enquire whether Bacon's actual writings have anything relevant to link with the present collection. However, enquiries soon reveal

Conclusions

that, in attempting to apply, insofar as it is feasible, 'scientific method' to the investigation of phenomena in their natural setting, we have re-created Bacon's own style of enquiry to a startling degree. For instance, to Bacon "It was reported by a sober man, that an artificial spring may be made thus: Find out a hanging ground, where there is a good quick fall of rainwater. Lay a half-trough of stone, of good length, three or four feet deep within the same ground; with one end upon the high ground, the other upon the low. Cover the trough with brakes a good thickness, and cast sand upon the top of the brakes: you shall see, saith he, that after some showers are past, the lower end of the trough will run like a spring of water: which is no marvel, if it hold while the rain water lasteth; but he said it would continue long time after the rain is past" (Montagu 1826, page 13). Not only is 'interflow' precisely described in this quotation, but so also is the need to "find out" a particular location. This sense of the significance of location is repeatedly in evidence. For example "You may make a judgement of waters according to the place whence they spring or come ... Springs on the tops of high hills are the best... For waters in valleys join in effect under ground with all waters of the same level; whereas springs in the tops of hills pass through a great deal of pure earth with less mixture of other waters ... Neither may you trust water that tastes sweet, for they are commonly found in rising grounds of great cities, which must needs take in a great deal of filth" (Montagu 1826, pages 194-196). He also appreciated links between localities. " ... if you take earth from land adjoining the river Nile, and preserve it in that manner that it neither comes to be wet nor wasted; and weigh it daily, it will not alter weight until the seventeenth of June, which is the day when the river beginneth to rise; and then will grow more and more ponderous, till the river cometh to its height ... upon that day when the river first riseth, great plagues in Cairo use suddenly to break up" (Montagu 1826, page 392). Bacon's acute awareness of seasonal change included an awareness of time lags. He reported that " ... water in wells is warmer in winter than in summer; and so air in caves" (Montagu 1826, page 478). There is also the substitution of time for places, and place for time. " ... for cold we must stay till it cometh, or seek it in deep caves, or high mountains: and when all is done, we cannot obtain it in any great degree: for furnaces of fire are far hotter than a summer's sun; but vaults or hills are not much colder than a winter's frost" (Montagu 1826, page 45).

More generally, Bacon made considerable use of the expanding geographical knowledge of the time. Exploration and discovery gave him one of the most powerful arguments for arousing dissatisfaction with " ... the restrictions imposed by traditional dogmatists, emphasising the rights of ancient authors ... The ancients gained a knowledge of but a small portion of the world. The travels of Pythagoras, Democritus, and Plato were suburban jaunts

rather than journeys afar. Many places and climates which
they described as uninhabitable have been found to be
populated with people of varied customs and lives"
(Anderson 1948, page 115). Even maps were used to support
his arguments. "Scientists in their theorizing would do
well ... to avoid excessive generalities which give no
traceable information about things and are about as useful
in practice as is an Ortelius' universal map to direct the
way between London and York" (Anderson 1948, page 101-102).

If the actual behaviour of environmental phenomena
added substance to Bacon's eloquent exhortations, several
significant reasons might be suggested to explain why the
more detailed pursuit of such studies did not become part
of the science which took shape after his death. First,
and although he avowed " ... that the legitimate interpretation of nature should in the first ascent, before a
certain stage of generality is reached, be kept pure and
segregate from all application to works ..." (Anderson
1948, page 11), his was a much-needed materialist philosophy that would " ... create manifold works for the
relief of man's estate". Secondly, he was clearly
exasperated by Nature not handing to him sufficient degrees
of range in temperature. "As for heat, many varities of
this are available ... gentle heats, fierce heats,
regular heats; increases and decreases of heat"
(Anderson 1948, page 255). In contrast, "We are confined
in experiment by cold mainly to wintry frost, cold of caves,
and cold made by surrounding bodies with snow and ice ...
Since Nature applies cold sparingly, we must look for
substitutes for cold" (Anderson 1948, pages 154-255). The
need for a laboratory environment and control was obvious
and became axiomatic after Bacon caught his death of cold
when stuffing a chicken with snow. "Whosoever will be an
inquirer into nature, let him resort to a conservatory of
snow and ice, such as they use for delicacy to cool wine
in summer" (Montagu 1826, page 46). A third, critical
fact, was that Bacon was opposed to a distinction between
artificial and natural phenomena. He demanded that it
" ... be firmly settled within the minds of men, that the
artificial does not differ from the natural either in form
or in essence" (Anderson 1948, page 202). Thus, in
envisaging a new sort of learned foundation, the College of
the Six Days' Works, he foresaw not just engines for
manufacturing heats of all sorts and degrees but also the
transference of all nature to within its walls. Botanical
gardens are provided for the study of certain herbs. Rare
beasts are available for study in parks and other enclosures.
Also envisaged are " ... Lakes for transferring of water from
a fresh or a salt condition into its respective opposite.
Cataracts and streams aid in the study of lively motions.
Means are provided for the creation of artificial snow,
hail, rain, thunder, and lightning" (Anderson 1948, page 24).
Fourthly, a biographical factor is critical. Coupled with
his weak constitution was his upbringing as a courtier.
His preference was for worldly magnificence and a princely
life-style. In connection with plans for the learned
foundation, Rawley wrote " ... I have heard his lordship

speak complainingly, that his lordship (who thinketh he deserveth to be an architect in this building) should be forced to be a workman and a labourer, and to dig clay and burn brick ...". Bacon, to his bitterly ironical end, was not a "field man".

2. Agricultural experiments and statistical methods

Few followed Bacon's exhortations that " ... everything respecting natural bodies and virtues be, as far as possible, numbered, weighed, measured and defined" more assiduously than agricultural experimentalist, John H. Gilbert. Born in Hull in 1817, and in part trained in Liebig's laboratory at Giessen, John Gilbert proceeded to join gentleman farmer John Lawes at his Rothamsted estate in 1843. There then started a 57-year partnership which amassed data from investigations so numerous, varied, and long-continued that " ... a library would scarce be sufficient to chronicle" (Tipper 1897). When E. J. Russell took charge at Rothamsted, he found there "great files of records" and R. A. Fisher was called in to examine " ... whether they were suitable for proper statistical examination and might be expected to yield more information than we had extracted. He revolutionised many of our ways of thinking out our research programmes, and in particular our methods of doing field experiments" (Russell 1966, page 326).

Two of the main conclusions with R. A. Fisher came to from an examination of the Lawes-Gilbert monuments of data are particularly relevant to any investigation of environmental phenomena in their natural setting. There was an emphatic challenge to 'Baconian principles', since some of Fisher's suggestions " ... ran counter to the general view then held that scientific investigation should proceed by asking one question at a time", and came in for much criticism (Russell 1966, page 379). Fisher's view, initially stated in 1935, was that "In expositions of the scientific use of experimentation it is frequent to find an excessive stress laid on the importance of varying the essential conditions *only one at a time* (his italics) (Fisher 1971, page 93). He did not agree that to do so is "an ideal scientific procedure" because in genuine research " ... we are usually ignorant which, out of innumerable possible factors, may prove ultimately to be the most important" (Fisher 1971, pages 93-94). Furthermore, "Whatever degree of care and experimental skill is expended in equalising the conditions, other than the one under test, which are liable to affect the result, this equalisation must always be to a greater or less extent incomplete" (Fisher 1971, page 19). The statistical methods devised by Fisher and others have proved indispensable aids to all those wishing to enquire into the nature of relationships involving the simultaneous operation of several variables and for studies in the natural environment, where laboratory controls cannot be achieved. However, the laws of chance which are the exclusive controls on experiments described by Fisher, depending on a strict randomising of the experiments' arrangements, represent a

Conclusions

return to the inescapable heterogeneity of the earth's surface. In R. A. Fisher's book *The design of experiments* few phrases appear more often than does 'soil heterogeneity' and one subsection is entitled "The problems of controlling heterogeneity". Beyond the fences and hedges of an agricultural experimental station, however, studies of inter-related environmental variables are much less readily randomised. Indeed, what is described in the present volume as a 'geographical approach' is often the very anthesis of randomisation, in the sense that a certain type of locality is defined and deliberately sought. Since statistical methods have been one of the most recent possible 'unifying' themes suggested for geography, it is not easy to argue, in a sentence, that randomization inevitably blocks out the vista from a geographical viewpoint. In the present collection, instead of perceiving 'heterogeneity' as a problem requiring homogenisation or randomisation, but a characteristic to be *utilized,* two important features minimize the clash with statistics which depend on a valid estimate of error. First, in most investigations the studies are not based on a sample but on such a number of cases as to approach the universe of phenomena under investigation. Thus, virtually *all* outcrops are studied with the limestone hill studied in Malaya (Chapter 3), virtually all the springs in Chapel-le-Dale are sampled (Chapter 4), all catchments in the North York Moors are sampled (Chapter 8), and the coverage of long profiles leaves few excepted. In other cases, the number of observations is very large, with number of springs (Chapter 10) or measurements of meander dimensions (Chapter 13) running into hundreds. Secondly, the 'significance' of such observations is to be evaluated, as are departures from general trends, in the light of knowledge of the specific characteristics of the locations in question rather than in the light of statistical tests for significance which depend on randomisation, giving every locality an equal change of selection. In unexplored areas of geomorphology, randomisation is particularly important in ensuring a balanced and impartial view of the phenomena under investigation (Pitty 1966). In tackling familiar problems, however, implicit in an insistence on randomization would be tacitly to assume less knowledge or more ignorance of the objects of interest than is, in fact, the case; " ... an erroneous assumption of ignorance is not innocuous; it often leads to manifest absurdities. Experimenters should remember that they and their colleagues usually know more about the kind of material they are dealing with than do the authors of text-books written without such personal experience" (Fisher 1971, page 49).

SOME GENERAL CONCLUSIONS
AND DEFINITIONS

Since geomorphology, unlike geography, has a specific object at the centre of its enquiries, those described or self-styled as geomorphologists are spared the difficulties of geographers who become concerned to define an area of study specific to geography. Geomorphologists, whatever

Conclusions

may be the range or even clashing shades of interest that lie within in any coherent field of study, know that in focussing their attention on the shape of the landsurface and the processes involved in producing erosional and depositional shapes on that surface, theirs is a distinctive pursuit. Furthermore, patterns of thought in geomorphology are more easily traced backwards than in geography, since the specific object of study identifies all who have described and examined landforms, regardless of their professional role or personal interests, as contributors to geomorphology. In contrast, any strand of methodological enquiry into the nature and purpose of geography could prove to be simply any one of a dozen loose ends, only one of which may end up on the North European Plain, some 100 or 150 years ago. Nonetheless, many striking parallels with Hartshorne's views have been found and some exact similarities identified. Therefore, as well as hoping to have made some contribution to that important aspect of geomorphology concerned with fluvial forms and processes, the claim to have adopted a 'geographical' approach seems, in general, justified. In detail perhaps the main contrast is the view of heterogeneity at the earth's surface. For Hartshorne this was something initially to be regionalised, compartmentalised into homogeneous units. Later Hartshorne (1959) seemed disillusioned with this approach, referring repeatedly to the need to 'break down' areal differentiations. For R. A. Fisher and the statistical methodologies which followed, heterogeneity was a problem, too much the source of unequal chance occurrences, requiring randomization at the outset. Although centuries earlier Francis Bacon had grasped the significance of variability in the environment and the use to which both this and temporal change within a season could be put, he was oblivious of the fact that a change of location does, for most environmental phenomena, mean some intrinsic change in their character.

Unlike the above approaches, exemplified in the present book is the deliberation given to the common, if only intuitive, practice of searching for suitable places in which to settle one's investigations. Instead of homogenizing, regionalizing, breaking down, randomising or extracting variables from their natural context, the emphasis is on *utilizing* areal differentiation. However, in adopting naturally simplified areas it would be unwise to suggest that 'control' has been introduced. As Fisher warned, "The authoritative assertion 'His *controls* are *totally* inadequate" must have temporarily discredited many a promising line of work" (Fisher 1971, page 2). On the other hand, the procedure relies too heavily on the investigators personal experience and decisions for statistical terms like 'sample' and 'significance' to be appropriate. Therefore a second key word, in addition to *utilization* is needed to describe a midway point between laboratory 'control' and statistical 'variable'. Provisionally the word 'selector' or phrase 'natural selector' is adopted to describe an environmental characteristic which a deliberate choice of study area maximises or minimizes.

278

Conclusions

Therefore, the 'geographical approaches' in the present book might support the following, tentative, definition. "Geographical method *utilizes* the variability at the earth's surface as a source of natural selectors in its investigations, and depends on the exploration for, and discovery of, localities naturally simplified by the marked presence or absence of selectors of particular interest".

REFERENCES

Anderson, F.H., 1948. *The philosophy of Francis Bacon.* (University of Chicago Press, Chicago)

Appleton, J., 1975. *The experience of landscape.* (Wiley, London)

Baker, J. N. L., 1937. *A history of geographical discovery and exploration.* (Harrap, London)

Beaglehole, J.C., 1966. *The exploration of the Pacific.* (Black, London), 3rd edition

Berry, B.J.L., 1964. Approaches to regional analysis: a synthesis. *Annals of the Association of American Geographers,* 54, 2-11

Booker, C., 1969. *The Neophiliacs.* (Collins, London)

Courtenay, P.P., 1979. The origin of the title "James Cook University of North Queensland'. *(personal communication to M. Bonell)*

Dickinson, R.E., 1939. Landscape and society. *The Scottish Geographical Magazine,* 55, 1-4

Fisher, R.A., 1971. *The design of experiments.* (Hafner, New York), 8th edition

Harris, J.W.K. & Herbich, I., 1978. Aspects of early Pleistocene hominid behaviour east of Lake Turkana. In: *Geological background to fossil man,* ed. W. W. Bishop (Scottish Academic Press, Edinburgh), 529-547

Hartshorne, R., 1939. *The nature of geography: a critical survey of current thought in the light of the past.* (Association of American Geographers, Lancaster, Pennsylvania)

Hartshorne, R., 1959. *Perspectives on the nature of geography.* (Rand McNally, Chicago)

Herbertson, A.J., 1916. Regional environment, heredity and consciousness. *The Geographical Teacher,* 8, 147-153

Huntington, E., 1945. *Mainsprings of civilisation.* (Wiley, New York)

James, P.E., 1934. The terminology of regional description. *Annals of the Association of American Geographers,* 24, 77-86

Johanson, D.C., 1976. Ethiopia yields first 'family' of early man. *The National Geographic Magazine,* 150, 790-811

Conclusions

Leighly, J., 1955. What has happened to physical geography? *Annals of the Association of American Geographers*, 45, 309-318

Leopold, L.B., Wolman, M.G. & Miller, J.P., 1964. *Fluvial processes in geomorphology*. (Freeman, San Francisco)

Montagu, B., 1826. *The works of Francis Bacon, Lord Chancellor of England*. (Pickering, London), Volume IV

Morgan, E., 1976. *Falling apart: the rise and decline of urban civilization*. (Souvenir Press, London)

Pitty, A.F., 1966. Some problems in the location and delimitation of slope-profiles. *Zeitschrift für Geomorphologie*, 10, 454-461

Pontefract, E. & Hartley, M., 1938. *Wharfedale*. (Dent, London)

Russell, E.J., 1966. *A history of agricultural science in Great Britain 1620-1954*. (Allen & Unwin, London)

Sollas, W.J. et al., 1897. Report of the Committee ... to investigate the recent bog-flow in Kerry. *Scientific Proceedings of the Royal Dublin Society*, 8, 475-507

Taylor, T.G., 1951. *Geography in the twentieth century: a study of growth, field techniques, aims and trends*. (Methuen, London)

Tipper, C.J.R., 1897. *The Rothamsted experiments and their practical lessons for farmers*. (Crosby Lockwood, London)

Whitaker, V.K., 1962. *Francis Bacon's intellectual milieu*. (University of California Library, Los Angeles)

Wooldridge, S.W., 1951. Some reflections on the role and relations of Geomprohology. In: *London essays in Geography*, eds. L. D. Stamp & S. W. Wooldridge (Longman, Green, London), 19-31

Wooldridge, S.W. & Kircaldy, J.F., 1936. River profiles and denudation chronology in southern England. *The Geological Magazine*, 73, 1-6

Wooldridge, S.W. & Linton, D.L., 1955. *Structure, surface and drainage in south-east England*. (Philip, London)

Wrigley, E.A., 1965. Changes in the philosophy of geography. In: *Frontiers in geographical teaching*, eds. R. J. Chorley & P. Haggett. (Methuen, London), 3-20

AUTHOR INDEX

Acquaye, D. 24
Agar, R. 257
Allen, J.R.L. 179, 190, 192
Anderson, F.H. 273, 275
Anderson, G.D. 114
Anderson, H.W. 128
Arnett, R.R. 127, 143
Asfari, A.F. 173
Atkinson, F. 60
Atterberg, A. 114-115
Axelsson, V. 190

Bailey, S.W. 75
Baker, V.R. 228
Barclay-Estrup, P. 143
Bathurst, J.C. 173
Bauer, F. 38
Bauer, L. 128
Baver, L.D. 115
Bellamy, D.J. 149, 157
Bendelow, V.C. 113
Bennison, G.M. 251
Berhard, F. 75
Berry, B.J.L. 271
Bishopp, D.W. 157
Black, P.E. 193
Blackie, J.R. 73
Boatman, D.J. 155
Boelter, D.H. 156
Bonell, M. 2, 81-84, 87-88
Booker, C. 264
Bormann, F.H. 6, 127
Boswell, K.C. 242
Bott, M.H.P. 53, 55
Bowen, B.E. 182, 196
Bower, M.M. 160
Bowser, R. 62
Bradley, W.C. 208
Brammall, A. 6
Bray, J.R. 75
Bray, L.G. 41
Brice, J.C. 223, 227
Bridge, J.S. 173, 192
Bridges, E.M. 128
Briggs, D.J. 121
Brindle, B.J. 228
Brinkman, R. 8
British Standards Institution 116
Brook, D. 61
Brown, A.H.F. 14
Brown, E.H. 241-242

Brunsden, D. 8, 13
Brush, L.M. 196
Burges, A. 6
Butzer, K.W. 182

Cailleux, A. 203-204
Carlisle, A. 14
Carlston, C.W. 216, 235
Carson, M.A. 173
Carter, W.L. 62
Carroll, D.M. 113
Catt, J.A. 114
Cavanagh, A.H. 64
Chapman, G. 76
Chapman, S.B. 156
Chitale, S.W. 235
Chow, V.T. 193
Christ, C.L. 3.
Clapham, A.R. 153
Clayden, B. 11
Clayton, K.M. 250
Coase, A. 59
Colby, B.R. 188
Coleman, J.M. 179
Colhoun, E. 157-158
Cooper, R.G. 2, 121-122
Courtenay, P.P. 272
Courtney, F.M. 121
Crickmore, M.J. 188
Crowther, J. 31-32
Cryer, R. 134
Culbertson, J.K. 173

Dagg, M. 73
Dakyns, J.R. 51
Dal Cin, R. 211
Daniels, R.E. 156
Delap, A.D. 157
Diaconu, C. 128
Dickinson, R.E. 264
Dickson, F.H. 190
Diskin, M.N. 182
Doughty, P.S. 55
Douglas, I. 73, 76, 94, 128, 196
Drake, W.D. 39
Dunham, K.C. 53, 157, 251
Dunne, T. 73
Dury, G.H. 168, 227, 236
Dwerryhouse, A.R. 62

Einstein, H.A. 173
Embleton, C. 255
Evans, O.F. 192

Ewel, J.J. 77
Exley, C.S. 6

Fahnestock, R.K. 192
Ferguson, R.I. 227, 235
Findlater, I.F. 182
Fisher, R.A. 276-278
Fitch, F.J. 184
Fitton, E.P. 121
Folk, R.L. 116, 208
Fournier, R. 127-128
Fox, W.T. 191
Fox-Strangways, C. 110
Francis, J.R.D. 173
Frederickson, A.F. 8
Frostick, L.E. 2, 179, 184, 192

Garrels, R.M. 31
Garwood, E.J. 51
Gascoyne, M. 58
George, T.N. 250, 255
Gill, W.R. 123
Gilmour, D.A. 2, 81-83, 87-88
Gimingham, C.H. 60-61
Glymph, L.M. 128
Gobbett, D.J. 32
Godwin, H. 247
Goh, K.C. 73
Goode, D.A. 153
Goodyear, E. 51
Gorham, E. 75
Grainger, B.M. 60
Gregory, K.J. 254
Gresswell, R.K. 247

Hack, J.T. 188, 196
Haig, I.T. 76
Hall, D.G. 132
Halliwell, R.A. 51-52
Hanshaw, B.B. 39
Harding, D.M. 128
Harmer, F.W. 250
Harms, J.C. 192
Harris, J.W.K. 269
Harrod, T.R. 11
Hartley, M. 268
Hartshorne, R. 263-265, 267, 270-271, 278
Harvey, A.M. 128, 221
Harwood, H.F. 6
Hemingway, J.E. 61, 128
Henderson, M.R. 35
Herbertson, A.J. 266
Herbich, I. 269
Hewlett, J.D. 73, 82, 87
Hibbert, A.R. 73, 87

High, L.R. 192
Hillel, D. 76
Hodgson, J.M. 113
Holeman, J.N. 127
Hollingworth, S.E. 247
Hooke, R. LeB. 173
Horton, R.E. 73, 163, 167-168
Hosking, J.R. 209
Huddleston, F. 157-158
Hudson, N. 74, 93
Hudson, R.G.S. 204
Huntington, E. 266

Imeson, A.C. 2, 94, 99, 102, 114, 127, 142, 160
Ingram, H.A.P. 158
Isbell, R.F. 76

Jackson, I.K. 75, 76
James, P.E. 266
James, P.W. 153
Jansen, J.M. 128
Johanson, D.C. 269
Johnson, G.A.L. 157
Jones, C.R. 32
Jones, O.T. 242
Jordan, C.F. 77
Jowsey, P.C. 149
Jungerius, P.D. 94
Jarvis, J. 173
Jennings, J.N. 251
Jopling, A.V.

Keller, E.A. 220, 229
Kennedy, V.C. 188
Kendall, P.F. 254
Kent, M. 11
Kenworthy, J.B. 73
Keppel, R.V. 184
Kesel, R.H. 76
Kidson, C. 255
King, C.A.M. 51
King, W.B.R. 253-254
Kircaldy, J.F. 269
Kline, J.R. 77
Klinge, H. 75
Klingeman, P.C. 193
Knapp, B.J. 82
Knighton, A.D. 229
Kolmer, J.R. 191
Kouba, D.L. 188
Krumbein, W.C. 203, 206
Kuenen, Ph.H. 192, 209
Kwaad, F.J.P.M. 94, 96, 99, 100

Lack, T.J. 24
Lane, B.E. 14
Lane, L.J. 182
Langbein, W.B. 241-243
Lapworth, C. 250
Laronne, J.B. 173
Leaf, C.F. 94
Leakey, R.E.F. 180
Ledger, D.L. 73
Leighly, J. 269
Leopold, L.B. 121, 127, 179, 184, 196, 204, 216, 228, 235, 241, 264, 267-268
Lewin, J. 228
Lewis, W.V. 241
Lieblien, J. 203
Likens, G.E. 127
Linton, D.L. 258, 270
Lockwood, J.G. 75
Long, M.H. 62
Loughnan, F.C. 24
Lovegrove, E.J. 209-210
Low, K.S. 73

Mabesoone, J.M. 204
Mackin, J.H. 241, 249
Madge, D.S. 77
Maner, S.B. 128
Manley, D.J.R. 11
Martinec, I.J. 173
Mathews, W.H. 190
Mayfield, B. 159
McBride, E.F. 192
McCammon, R.B. 204
McKee, E.D. 192
Meland, N. 173
Meteorological Office, 111
Miller, J.P. 179, 184, 203, 264, 267-268
Mina, V.N. 24
Ministry of Transport, 209
Mississippi River Commission 186
Mitchell, D. 121
Mitchell, G.F. 157
Montagu, B. 274-275
Moore, P.D. 149-157
Morgan, E. 267
Morgan, R.S. 250
Morris, P.G. 168
Moss, A.J. 19
Müller, R. 179, 190
Myers, J.O. 60

Naidu, A.S. 186
Normark, W.R. 190

Norrman, J.O. 173
Nutter, J.D. 73
Nye, P.H. 76

O'Nions, R.K. 53
Ovenshine, A.T. 208
Ouma, J.P.B.M. 204

Painter, R.B. 128
Palmer, A. 56
Palmer, J. 121, 124
Parizek, R.R. 14
Paton, J.R. 33
Pearson, M.C. 156, 159
Pereira, H.C. 73
Pettijohn, F.J. 192
Phemister, J. 250
Phillips, J. 204
Picard, M.D. 192
Picknett, R.G. 31
Pickup, G. 221
Piest, R.F. 133
Pittman, E.D. 208
Pitty, A.F. 63, 114, 121, 165-166, 168, 277
Plumley, W.J. 186, 208, 211
Plummer, L.M. 31
Pollack, J.M. 186
Pontefract, E. 268
Poole, E.G. 250
Potter, P.E. 186
Prentice, J.E. 172
Price, R.J. 221
Pringle, J. 150
Pugh, J.C. 221

Ramsbottom, W.H.C. 53, 61
Ramsden, R.W. 62
Ratcliffe, D.A. 150
Rayner, D. H. 128, 204
Reaves, C.A. 123
Reed, F.R.C. 250
Reid, I. 2, 179, 184, 192
Renard, K.G. 184
Richards, K.S. 228
Richards, P.W. 153
Rightmire, C.T. 39
Riley, S.J. 222
Ritter, D.F. 228
Roberts, M.C. 193
Robertson, R.A. 149, 156
Rodrigues, W. 75
Roques, H. 41
Rubey, W.W. 241-242
Rule, A. 60
Russell, E.J. 276
Rutter, A.J. 76

Ruxton, B.P.   73, 76
Rycroft, D.W.   156
Ryder, P.F.   56

Sandeman, E.   10, 13
Sanders, J.E.   192
Savigear, R.A.G.   121
Scheffer, R.   203
Schimper, A.F.W.   75
Schumm, S.A.   216, 226, 229, 231
Seed, H.B.   115
Sharon, D.   182
Shepard, F.P.   179, 190
Shreve, R.L.   163
Shulits, S.   241
Shuster, E.T.   63, 65
Siever, R.   24
Simmons, I.G.   11
Simpson, E.   60
Skempton, A.W.   115
Smith, C.J.   17
Smith, N.D.   173, 179, 192
Smith, R.M.   94
Sneed, E.D.   208
Soane, B.D.   123
Sollas, W.J.   157, 268
Stamey, W.L.   94
Stark, N.   76
Statham, O.W.   62, 64
Stone, M.   6
Strakhov, N.M.   127
Sweeting, G.S.   53
Sweeting, M.M.   47, 53

Tallis, J.H.   156, 157, 160
Taylor, T.G.   266
Ternan, J.L.   1, 65, 165, 170-171
Thomas, M.   73
Thomas, T.M.   160
Thornes, J.B.   184
Tiddeman, R.H.   60
Tille, W.   128
Tinsley, J.   24
Tipper, C.J.R.   276
Tomlinson, R.W.   155
Toy, T.J.   93
Tracey, G.   75
Tricart, J.   73, 203-204
Trudgill, S.T.   39
Truesdale, V.W.   17
Tubey, L.W.   209
Tukey, H.B.   5
Tutin, T.G.   153

van Zon, H.   2, 96, 100-101
Varnes, D.J.   120
Verstraten, J.M.   6, 97
Vogt, H.   203
Vondra, C.F.   182, 196

Wager, L.R.   55-56, 206
Walker, D.   150
Wallace, E.C.   153
Wallis, J.R.   128
Waltham, A.C.   53, 56-57, 60
Wanner, H.   75
Warburg, E.F.   153
Ward, R.C.   193
Ward, W.C.   116
Warner, R.F.   221
Warwick, G.T.   168
Waters, R.S.   8, 11
Watkins, R.T.   187, 195
Watt, A.S.   142
Webb, J.L.   75, 77
Wentworth, C.K.   203
West, E.A.   253
Weyman, D.R.   17
Wheeler, D.A.   2
Whetton, J.T.   188, 196
Whipkey, R.Z.   14
White, E.J.   14
White, W.B.   63, 65
Whiteman, A.J.   250
Whitmore, T.C.   73
Whittaker, E.   143
Whittel, P.A.   2, 170, 171, 177
Wigley, T.M.L.   31, 39
Wilcockson, W.H.   61
Williams, A.G.   1, 165
Williams, J.C.   247
Williams, R.C.   94
Wills, L.J.   247, 250
Wolman, W.G.   206, 216, 221, 235, 264, 267-268
Wood, W.W.   14
Wooding, R.A.   184
Wooldridge, S.W.   250, 269-270
Wright, A.E.   251
Wright, L.D.   179

Yen, B.C.   193

GEOGRAPHICAL INDEX

Aberystwyth 243
Africa 261
  East 73
  West 73
Aire, River 244, 249
Amazon forest 75-77
Anak Bukit Takum 31, 33-35,
  47-48
Antrim 157-158
Appalachians 88
Arctic-alpine environments
  75
Ardennes 93-94, 98
Arid zone 179, 182, 184,
  188, 196
  environment 184, 203
Arizona 184
Arroyo Seco, Los Angeles
  County 203
Ashberry 110, 121-122
Asia, South-east 128, 261
Askrigg block 53, 55-56,
  59, 204
Australia 77, 261, 264
  north-east 73
  northern 80
Ayr, River 244, 249, 255
Aysgarth Falls 253

Babinda 73-75, 77-78,
  80-81, 83, 89
Barbondale 59
Big Sandy River (Virginia)
  203
Black Hills, Dakota 208, 211
Black Water 217-220, 222, 231
Bodmin Moor 252
Bolton Bridge 166
Borneo 75
Bowmont Water 217, 219-220,
  227, 232
Brandywine Creek (Maryland)
  203
Bransdale 133
Brazil 204
Brishie Bog 150-155, 158,
  262
Britain 1, 58-59, 149, 157,
  160, 242, 252, 257
  Upland 215
British Isles 149, 204, 242-
  243, 246, 256-257, 261
Bristol Channel 250
Buxton 165

Cairns 83
Camel, River 244, 252
Canberra 251
Canterbury 268
Carsphairn Lane River
  217-220, 232
Castleton 164
Caydale 114
Chapel-le-Dale 52, 54-55,
  60-63, 65, 164, 277
Cheshire Plain 250
Clapham 61, 67
Clare, County 37, 157
Colne, River 243-244,
  249, 255-256
Colorado 94, 188
Columbia, River 188
Combshead Tor 9-10
Cooran Lane 150-151
County Clare 37, 157
County Donegal 157
County Mayo 157-158
County Wicklow 157

Dachstein Alps 38
Dart, River 244, 249, 252
Dartmoor 5-6, 10-11, 15,
  28, 252, 262, 268,
  271
Dee, River (North Wales)
  244, 247, 249-250
Deifenbaach catchment
  93-100
Derwent, River (Cumbria)
  243-244, 249
Derwent, River (Yorkshire)
  129-130, 132, 134,
  136, 138, 140, 166,
  215, 244, 246-247,
  249-251
Devil's Bridge 243
Dionard, River 244, 249-
  250
Donegal, County 157
Dorset Heights 246-247
Dove, River (Derbyshire)
  164, 166, 244
Dove, River (Yorkshire)
  129-130, 132, 136,
  138, 142, 166-167, 215,
  244
Durham 268

East Beck Row 129-130, 132, 134, 136, 138, 140
Ebbw, River 244, 249, 255
Eden
  Garden of 268
  River 244, 247, 249
  Vale of 247
Elwy, River 244, 249, 255
England 242, 247
  North-east 128
  Northern 216, 262
  North-west 128, 268
  South-east 256
  Southern 268
  South-west 8-9, 252
Equatorial regions 33
Esk, River (Yorkshire) 129-130, 132, 136, 138, 242, 244, 249, 254, 257
Eskdale 254
Ettelbrück 93, 95
Europe 128, 203, 261, 264
Exe, River 244, 249, 252

Fal, River 244, 252
Falls of Clyde 273
Farndale 133
Flassen Dale 120
Forge Valley 166
Fowey, River 244, 249, 252

Galloway 150-151
Gaping Gill 56, 58, 60-61
Giggleswick Scar 251
God's Bridge 54, 62, 66, 166-167, 253
Gowerdale 114
Gragareth 52
Grampians 242
Grassington 170-171, 204, 208
Greta, River 52, 166, 244, 249, 251, 253
Guatemala 77
Gunong Gajah-Tempurong massif 32
Gutland 93-94, 98
Guyana 76

Haarts catchment 93, 95-97, 99-100, 102-103
Hacking River (New South Wales) 204
Halter Burn 217, 219-220, 235
Hambleton Hills 109-112, 114, 120-124
Helmsley 109, 111, 129, 164

Hodge Beck 129-130, 132, 136, 138, 166-167, 244
Honsschlaed catchment 94-96, 100
Howardian Hills 250
Hull 276
Humber, River 247, 250
Hutton Beck 129-130, 132, 136, 138, 166-167

Ingleborough 51-56, 59, 61, 65
Ingleby Beck 129-130, 132-134, 136, 138
Ingleton 53, 56, 64, 67, 251
Ironbridge Gorge 250
Italy 203
Ivory Coast 75

Java 75

Kennett, River 244, 249 255-257
Kenya, Northern 179, 262
Kepong 33, 47
Killarny 157
Kingledoors Burn 216-220, 231, 235
Kingsdale 52, 61-63
Koobi Fora 179-180, 184, 188, 191-192, 196
Kuala Lumpur 31, 33, 75

Lake District 250
'Lake Lapworth' 250
'Lake Pickering' 250
Lake Turkana 179-181, 191, 269
Lapland 38
Larochette 93
Leven, River 129-138, 244
Levisham Beck 129-130, 132-134, 136, 138, 140
Littondale 170
London 275
Lower Colorado, River 208
Luxembourg 93-96, 98, 262, 268

Malay Peninsula 32
Malaysia, West 31-33, 262-263, 277
Malham 65, 166
Manifold, River 164, 166-169
Mayo, County 157-158
Mendip 59

Merced, River (Sierra Nevada) 208
Mersey, River 244, 247, 249
Mississippi, River 188, 267

Naitiwa, River 185
Narrator Brook 5-11, 13, 15, 17, 28, 165
Naver, River 244, 249-250
Neath, River 244, 255
Nettledale 114
New Guinea 76
New Mexico 184, 203
Nidderdale 59
Nile, River 242, 274
North
  Carolina 94
  European Plain 278
  Sea 128, 134, 140, 250
  Yorkshire Moors 128-129, 139, 143, 164-166, 169, 215, 246, 262-263, 277
Norwich 269

Oklahoma 94
Omo, River 182
Oregon 94
Ouse, River 244, 247

Paris Basin 93
Parrett, River 244, 246-247, 249
Peak District 164-166, 169
Pennines 157
  Northern 55
  Southern 156, 164, 166, 268
Periglacial environments 5
Pickering Beck 129-130, 132, 136, 138, 166
Pickering, Lake 250
Pickering, Vale of 128, 166-167, 169, 215
Polmaddy Burn 217-219, 231
Puerto Rico 37

Queensland 73-75, 77-78, 94, 262

Ramsdale Beck 257
Red Hills 128
Rede, River 217, 219, 227
Rheidol, River 243
Rhymney, River 247
Ribble, River 60, 166, 244, 251

Riccal, River 129-130, 132, 134, 136, 138
Robin Hood's Bay 257
Rosedale 133-134
Rothamsted 276
Roughtor Plantation 10
Roumania 128
Rye, River 110, 114, 129-130, 133, 136, 138, 140, 167, 215, 244
Ryedale 110

San Gabriel Canyon 203, 206
Savanna 76-77
Scales Moor 52, 54, 57-58, 67
Scarborough 128, 164, 250
Scilly Isles 6
Scotland
  North-west Highlands 250
  Southern 152, 216, 221, 242
Scottish Highlands 258
Scugdale Beck 129-130, 132, 134, 136, 138
Sea Cut 129-130, 132, 136, 138
Semi-arid environments 5, 149, 203, 243, 262
Seven, River 129-130, 132, 136, 138, 142, 244
Severn, River 244, 247, 249, 250-251
Shrewsbury 250, 268
Silver Flowe 150
Skirfare, River 167, 170-171
Snaizeholme Beck 217, 219, 231, 236
Somerset
  Island (Canada) 38
  Levels 247
Sorrow Beck 129-130, 132, 136, 138
Southern Uplands 242
Stanhope Burn 217-220, 231
Swaledale 56-57
Swinsto Hole 52, 57, 59-60, 62

Taff, River 244, 247, 249, 255
Tamar, River 244, 252
Taw, River 244, 252
Tawe, River 244, 249, 255
Tees Plain 140
Teign, River 244, 249, 252
Temperate latitudes 3, 8, 73, 75-76, 88-89, 105, 246

Test, River   242, 244, 249
Texas   94
Thames, River   24, 249, 255
Thames Basin   256
Thuringa   128
Torridge, River   244, 252
Towy, River   242, 244, 255
Trent, River   244, 247, 249
Tropical areas   1-2, 73
  Humid   3, 8, 76, 78, 203
Tundra   11
Turkana, Lake   179-181, 191, 269

United States
  South-west   216, 221
Ure, River   217, 219, 231, 244, 253
Usk   244, 247

Vrynwy, River   250

Wabash Valley   204
Wales   242-243
  North   250
  South   59, 247
Wash, The   247
Wensleydale   53, 204, 206, 253
Westerdale Moor   142
Wharfe, River   166, 171, 204, 206-208, 244, 262
Wharfedale   56, 164, 170, 204-206, 209
White Limestone area (Jamaica)   38-39
Wicklow, County   157
Wiltz   93, 95
Winchester   268
Wye, River (Derbyshire)   164-166, 168-169

York   110-111, 128, 140, 275
  Vale of   128, 140, 247
Yorkshire
  North   109, 257
  North-east   110
  North-west   5, 47, 51-53, 58-59, 164-167, 251, 262
Ystwyth, River   242, 244
Yugoslavia   54

SUBJECT INDEX

Abrasion, 203-204, 208-209
Acacias 187
Accretion 192
Accumulators 24
Activity, Clay 115-116
Afforestation 11, 120
Aggradation 249-250
Agricultural
  areas 96
  crops 96
  engineers 93
  experiments 276-277
  practices 136, 140
  soils 96
Agrostis
  setacea 11
  tenuis 13, 14, 24
Air masses, Maritime 9
Alkalinity 210
Alluvial plains 31-32, 267
Alluvium 76, 110, 114, 173-178, 262
Altitude 61, 111, 117, 130, 132, 246-247, 252, 256, 274
Alumina 76
Alycaeus 43
Animal activity 98-99, 102-103, 262, 275
Anions 37
Aquifers 163, 167
Arroyos 184
Aspect 33-34, 142
Atmosphere 5, 63, 97, 265
Atmospheric
  factors 160, 166
  fallout 5
  pollution 150
Atterberg limits 114-116
Auxiliary outlets,
  Groundwater 167-168

Bacon, Francis 273-276, 278
Badlands 181-182
Bank
  erosion 235
  storage 167
Bark 24
Basal conglomerate 53, 56
Baseflow 5, 64, 87, 163, 167, 169, 188
Base-level 251, 257

Beach
  bars 179, 185-186, 191-192, 194
  lagoon 179, 185-186, 191
Bedforms 173, 182-184, 192
Bedload 193, 206, 211, 268
  materials 229, 235-236, 249
  micro-topography 173, 179, 195
  mobility 221
  size 215, 220-221, 228
  sorting 204, 220
  transport 173, 187
Bicarbonate 37, 62
Biogeochemical processes 6
Biological
  activity, Soil 11
  cycling 5
  phenomena 262
Biosphere 8
Biotite 8, 11
Blockfields 8
Block glide 120, 124
Boea spp. 35
Bog
  bursts 157-158, 268
  pools, 151, 153-156
  Raised 151-154, 156, and see Peat
Borehole 68, 268
Bottles, Sampling 14, 36-37
Boulder 220, 228, 269
  clay 60, 129, 133, 137, 140
Boundary friction 188
Bracken 9-11, 21, 24
  canopy 21
  fronds 24
  transect 13-14, 18, 20-21
Brown earths
  Acid 11, 13
  Gleyed 114
Bryophytes 13, and see Mosses
Burning 114, 143, 160
Burrowing 98-100, 102-103, 106
Calcareous
  cement 53
  rocks 133-134, 136, 140
Calcicole species 35
Calcite 31, 46, 58, 63, 209
Calcium 35-36, 62, 64, 184
  carbonate 32, 36, 62, 66, and see Hardness

Calluna 14, 144
  vulgaris 142
Callunetum 114, 142
Caprock 63, 112, 204, 206
Carbonates 116
Carbon dioxide 6, 26, 31,
    39, 41, 44, 46, 66
Carbonic acid 6
Carex spp. 159
Cations 36, 41, 115
Cave 164-165, 265, 274-275
  collapse 58
  passages 56-58, 60
  systems 55-59, 63
  vadose 59
Chalk 249, 255-256
Channel
  bank 97, 99, 103, 211,
    220-221, 228
    collapse 229
    strength 229, 231-232,
      235
  bars 173, 179, 228
  cross section 129, 215-
    216, 220-226, 236
    area 224, 226, 229,
      235-236
    asymmetry 222, 226,
      229-233, 235
    depth 221, 224-228,
      232-234, 236
    perimeter 221, 224, 226-
      227, 236
    width 188-189, 216, 221-
      222, 228, 232-234,
      236
      Bankfull 216, 221
      -depth ratio 220, 229,
        231, 233
      relationships 228-229,
        233
  density 138, 140, 182
  deposits 188, 208
  Dry 167, 206
  ephemeral 184, 187, 206
  fill 179, 185-192, 211,
    228
  floor 206, 221
  geometry, also see Hyd-
    raulic geometry
  gorge 254
  gradient 100, 104, 184,
    189, 215, 220, 228,
    231, 243, 246
  head 163
  incision 98, 255
  length 137, 188
  migration 185, 228-229,
    231-233, 235, 237

  morphology 2
  network 163
  Paleo- 229
  plan-pattern, 2, 89, 127,
    173, 190, 215-216, 236,
    263, 270
    angle-of-turn 223-224,
      226, 236
    curvature 22, 226, 235-
      236
    direction change 223-224,
      226, 236
    sinuosity 215, 220, 226-
      228, 231-232, 235
    turn 228, 230, 235, 237
  pools 229, 232
  Relict 221
  Sand-bed 179, 215-216,
    221, 229, 232, 236
  sediment 102-106, 109,
    179-196, 225
  storage 184
  Subsurface 158-159
  Unconfined 215-216
  precipitation 103
Chemical reaction 31
Chirita spp. 35
Chlorite 11
Cladonia uncialis 153
Clay 79, 96, 110, 113-116,
  123, 172, 221, 228, 276
  formation 8
  mineralogy 11, 28, 115
  subsoil 59
Cliffs 33, 120, 257, also
  see Scars
Climate 9, 33-34, 73-77, 94
  109, 111, 113, 150, 157
Climatic
  changes 157, 160, 168-169,
    172
  zone 127-128
Coal Measures 56
Coastal recession 257
Cohesion 184, 220-221, 228
Colluvium 114, 182
Comminution 179
Competence 189, 191, 228
Condensation
Cook, James 272
Corallian rocks 110, 116, 121
Coriolis force 75
Craven Faults 55-56, 62, 67,
  171, 204, 206, 208, 251
Cremnophytes 35
Cross-bedding 192
Cultivation 93, 133
  arable 136-137

Cycle of erosion 8
Cyclone 75, 77
  'Keith' 86
  'Otto' 80
Cyclothems 51, 206
Cyperaceae 157

Data 1, 3, 31, 73, 94, 137, 273, 276
Decomposition, Organic matter 39, 75, 77-78
Deer 98, 101
Deforestation 11
Delta 179, 185, 190-194
Deltaic sedimentation 189, 196
Density-relief index 140, 143
Denudation 2, 23, 73, 127, 135, 137, 143-144, 252, 263
  chronology 8
Depth
  Flow 17, 82, 188
  Soil profile 26, 34, 88, 96
  weathering 10-11
Deschampsia flexuosa 114
Desert, Thorn-scrub 180
Diatom 23-24
Dilution 23, 26, 28, 64, 67, 133, 144, 204
Dip 57, 59, 182, 204, 206
Discharge 1, 22, 25, 63-65, 84, 88, 97, 102, 104, 106, 129, 132-133, 137, 142, 144, 168-169, 173, 188-189, 216, 234, 236, 243
  downstream increase 215
  -drainage area relationship 246
  Groundwater 163, also see Baseflow
  peak 103-104, 184
  pulses 184, 193, 195
Discordant streams 255
Discovery 192, 272-274, 289
Dissolved
  load 28, 144
  sediment concentration 127-128, 132-133, 136
  solid loss 137
Diurnal changes 156, 191
Dolomite 58, 63
Dolomitization 63
Downpour 2, 34, 43, 46, 74, 82, 123

Downwash 77
Drainage
  artificial 262
  basin 5, 127,-130, 217, 236
  analysis 127
  area 188, 243, 246, 263
  network 179, 182, 186, 191, 193, 195-196
  density 89, 129-130, 137, 140, 145
  ditches 97, 159
  Subsurface 158, 165, and see Interflow
  Underground 167, and see Cave
Drift 59, 134, 137, 215, 220, 250
Dualism 265, 272
Dunes 192
Durability, Bedload 2, 179, 220

Early Man's environment 196
Earth Science 3, 264
Ecology 3
Edaphic factor 76
Enclaves 169, 172
Environment, Natural 273-276
Environmental
  phenomena 278
  reconstruction 180
  science 3, 264
Environmentalism 266
Epiphytes 75
Equilibrium 241
Ericaceae 157
Erica tetralix 154
Eriophorum angustifolium 154
Erosion 8, 53, 77, 94, 128, 149
  Accelerated 143
  Gully 127, 166
  Peat 160
  pin 97
  rates 93, 105, 127, 262
  Splash 76, 94, 99-103, 105
Erosional loss 39
Estuarine lowlands 247
Estuary 242, 247
Evaporation 5, 46-47, 155-156, 160
Evapotranspiration,
  Potential 9, 18, 24, 113, 155

Experimental
  design 1, 3, 14, 35, 48, 262
  plot 99, 128, 142-144
Exploration 272-274, 289

Fault brecciation 58
Faulting 51-52, 55-58, 62
Feldspar 8
Felsite 208
Festuca-Agrostis grassland 114
Festuca ovina 13-14
Field
  capacity 115
  survey 231-232, 242
Fisher, R.A. 276-278
fissures 10, 13, 121-122, 265
Flood 25, 28, 58-59, 62, 64-66, 173
  bore 184, 187
  Catastrophic 268-269
  debris 182
  deposits 184-185, 188-189, 203, 206
  laminae 185
  discharge 133, 140, 167, 184-185
  event 179
  Flash 2, 185, 269-270
  generation 196
  Mean annual 217
  passage 183
  peak 163
    recession 186, 188
  -plain 79, 97, 127, 216, 228, 231
  recurrence interval 173, 196, 236
  stage 186
  water 192
    spread 188
  wave 102, 184, 188, 193
Flotsam 192
Flow
  components 163
  Concentrated 163
  Conduit- 63
  depth 42, 44, 46
  Diffuse- 63-65, 67
  Divergent 228
  duration 23
  Effective 221
  horizons 20
  Lateral 59-60, 81-84, 87, 156, and see Interflow

  length 39, 47-48
  Low 164
  magnitude 215, 234, 236
  path 78
  Plastic 120
  Return 97
  Subsurface 159, and see Interflow
  velocity 43-44, 46-47, 156, 188, 221, 237, 241
  volume 36, 42-43, 47, 164-165, 246
Fluid motion 173
Flushing 21, 26, 133
Flute, Solutional 31, 36
  gradient 40-43, 45, 47
  morphometry 40-41, 45
Folding 53, 55, 59
Forest 21, 24, 73, 75-76, 94, 96, 100, 105-106, 262, 265, 268
  areas 2, 88, 93, 96, 102
  canopy 76, 96, 98
  cyclone damage 77
  felling 93
  floor 75-78, 96, 101, 262, 265, 268, 270
  soils 96
  transect 13-14, 18, 20-21
Forestry 3
'Fossil' relief 54, 64-65
Fossils 191, 269
Fragipan 11, 13, 97
Freeze-thaw 94, 143, 166
Frost 8-9, 150, 274-275

Garcinia spp. 35
Geographers 264-267, 270-272, 277
Geography 1-3, 144, 261, 263-269, 277-278
  Regional 271
Geographical
  distribution 271
  location 65
  method 289
Geohydrology 55, 59, 110
Geological
  factors 31, 51, 252
  structural influences 57, 182, 184, 204, 251, 255, 262, and see Dip, Craven Faults, Faulting, Strike
  variations 3, 115, 128, 242, 261
Geologists 3, 51, 261, 267, 273
  American 261-264

Geology 3, 6, 32, 51-52,
         63, 93-94, 109, 114,
         182, 203, 217, 220,
         242, 255, 261
Gibbsite 28
Gilbert, John H. 276
Girvanella algal bed 51
Glacial
  activity 150, 241
  advance 169
  drift 220, and see Drift
  erosion 258
  infilling 247, 249
  materials 134
  maximum 169
  modification of drainage
    250, 254
  overdeepening 250
Gleying 114
Gneiss 76
Grade 258
Graded stream 241, 249
Gradient, and see Channel,
    Flute
  Cave passage 56
  Mire 160
  Longitudinal profile 241,
    255-257
Granite 1, 6, 10, 13, 53,
         80, 150, 165, 208,
         252-253, 262
  Decomposed 10, 13
  decomposition 6, 28
Grassland 9-11, 13-14, 18,
           20-21, 24
  transect 12-13, 18, 20
Gravel 2, 103-104, 189, 203
        206-207, 210-211,
        215, 228, 231, 236
Gravity anomaly 53
Grazing 11, 114
Grenzhorizont 157
Greywacke 53
Gritstone 51, 134, 204-
           206, 208, 210
Groundsurface 5, 98-99,
               122, 124, 163,
               262-263
Groundwater 5-6, 13, 63,
             103, 149, 151,
             163-164, 166-169,
             171-172, 187,
             210, 256, 265
Gully 156, 215, 220
  systems 128, 132-133,
           137, 140

Hardness
  alkaline 65-67
  calcium 36-38, 40, 63-65,
           67
  non-alkaline 36-41, 43-44
  total 36-47
Hartshorne, Richard 263-265,
    272, 278
Head deposits 8
Headwaters 28, 63, 78, 138,
            149, 160, 181, 188-190,
            193, 204, 209, 215, 220,
            242, 262
Heather 140, 142, and see
    Callunetum
  firing 11
Hercynian orogeny 6, 56
Heterogeneity, Earth-surface
    270-272, 277-279
Hettner, Alfred 264, 271
Human usage 150, 179, and
    see Agricultural, Man,
    Land-use
Humidity 77
Huntington, Ellsworth 266,
    269, 272
Hydraulic
  behaviour 192
  conductivity 82, 156-158,
                160
  equivalence 189
  geometry 173, 188-189, 227
  principles 235
  radius 226-227
  separation 179
  variable 196
Hydrogen ions 8
Hydrograph 9, 84, 87-88, 141,
            163, 167, 179, 184, 188,
            193, 196
Hydrological
  control 192
  pathways 5-6, 14, 20, 271
  processes 252
  variables 246-247
Hydrology 3
  Peatland 149, 154-155

Ice 133, 169, 250, 253, 275
  Devensian 250
  dispersal 150, 250
  Needle- 104
Impact 209
Impermeability
  Peat 156-157
  Rock 2, 51, 56, 110, 136,
        140, 171
  Subsoil 5-6, 17, 96-97

Induration 184
Infilling 247, 249-251
Infiltration 59, 168
  capacity 76, 140
  rate 76, 88, 96, 182
Ingletonian rocks 52-54, 61-62, 64, 251
Instability, Landsurface 2, 109, 120, 123, 133
  Peatland 160
Interception 35, 76, 106, 182
Interflow 6, 12, 14-18, 20-21, 103, 188, 274
  pipes 182
  wet-weather route 18, 26
Interglacial 168
Inter-tropical convergence zone (ITCZ) 75, 80
Intrusion 8, 53
Ions 6
Iron
  oxide 76
  pan 13, 18
  translocation 184
Irregularity, Rock surface 43

Joints 5, 8, 13, 32, 51, 53, 55-57, 59-60, 110, 120-121, 163, 167, 262
  Bedding 53, 55
Jurassic strata 93-94, 110, 112, 128, 130, 167

Kaolinite 8, 11, 28
Kaolinization 8
Karst features 54
Kellaways Rock 110, 112, 114
Kimmeridge Clay 128
Knick point 258, 268

Laboratory
  analysis 17, 36, 116, 129, 132, 186, 209, 221, 242-243
  conditions 31
  control 273
Lagg 151
Lagoon 191, 194
Lake 158, 166, 179, 185, 189, 196, 275
  ice marginal 254
  levels 182
  shore 182

Laminae, Horizontal parallel 192-193, 195-196
Landforms 6, 28, 137, 169, 215, 278
Landscape 3, 6, 62, 179, 264-266
Landsurface 127, 215, 241, 262, 278
Land-use 73, 96, 114, 128, 137, 140, 143, 145, 158-159, 268
Leaching 5, 24, 76-77
Leaf
  cover 103
  decay 5
  fall 99
  transport 100-101, 104
Leaves 5, 24, 104-105
Lichen 35, 40, 43, 153
Limestone 1, 31, 34, 37, 51-60, 63, 67, 164-165, 204, 206, 208, 262, 268, 277
  Great Scar 5, 51-55, 62, 204, 208
  Kingsdale 56
  Kuala Lumpur 32
  Oolitic 110, 113-114, 136
  pavements 47, 59, 62, 270
  pebbles 209
  Yoredale 51, 204
Liquid limit 113-117, 119, 123
Liquidity index 115, 117, 119-120
Lithology 51, 54, 93, 112, 128-130, 132, 137, 140, 145, 182, 206, 208, 210-211, 251-252, 261-262
Litter 2, 13-14, 34-35, 39, 48, 76, 86, 93, 96, 98-100, 105, 142, 262
  decomposition 6, 21, 99
  fall 75, 99
  layers 5-6, 18, 21, 77
  runoff 12, 16, 20
Littoral reworking 179
Loams 113, 116, 123
Location 274, 277
Loess 114
Longitudinal profile 2, 168, 184-185, 221, 241-258, 262-263, 270, 277
  concavity 184, 242-243, 246, 248-249
  index 242, 244, 248
  convexity 251-254
  dichotomy 253-254

discontinuity 254
extrapolation 241
irregularity 251, 254
'Over concave' 246-247, 249, 251-252
shape 241
smoothness 241
'Under concave' 246, 249, 251-254
Underground 59
Lower Lias 128

Magnesium 35-36, 62-63, 65, 67-68
Magnetite 195
Man, and see Agricultural, fossil 180, 262, 266-267, 271-272
  Mesolithic 11
  -Nature theme 266-269
Marble 32
Marl 94
Mass-movement 109-110, 120-121, 124, 262
Meandering 52, 94, 216, 226
Meanders 215, 268, 277, and see Channel plan-pattern
  'arc height' (or amplitude) 223, 226, 230, 233-237
  average bend radius 230, 235
  belt 227
  bend radius 223-224, 226
  planform 222, 230, 232, 235
  scale 215-216, 232, 235-236
  valley 168
  wavelength 216, 223, 233-236
Mechanical strength, Rock 33
Meltwater 133
Memecyclon spp. 35
Mesophyll Vine Forest 77
Metamorphic rocks 79, 150
  aureole 253
Metamorphism 32
Mica 11
Micrite 53, 60
Microgranite 8
Mineral
  ground 152, 158-159
  Heavy 186, 189, 192, 195
  inputs 150
  Light 186, 192
  particles 100, 103, 105, 142
Mineralogy 179
Mining 9

Mire 149-150, and see Peat
  Blanket 149
  'expanse' 154-155
  Raised 149
  Tertiary 149-150
  Valley 149
'Misfit' streams 227
Models 3, 135, 196, 261
Molinia 159
  caerule 11
Monophyllaea spp. 35
Moorland 9, 11, 13-14, 28, 114, 156, 166
Moraines 150
Moss 43, 153, 159, 210
Mudstones 55, 220

Nardus stricta 114
Natural Environment Research Council 205, 217, 236
Nature Reserve 150, 268
Nutrients 5-6, 11, 37, 75-77, 137, 151

Oak 11
Offshore fining 179, 196
Opaline particles 24
Organic
  acids 39, 41, 43, 46
  layer 142
  material 34, 59, 76, 116, 127, 195, 262, 265
  microbiological decomposition 39
Orthoclase 11
Outcrops 1, 5, 31, 37-43, 46-48, 103, 114, 215, 241, 262, 277
Overland flow 5, 73, 76, 78, 84, 86-88, 97, 101, 105, 142-143, 184
Oxford Clay 110, 112, 114, 116
Oxisols 76

Palaeoenvironmental reconstruction 196
Palaeogeography 196
Park 275
  National 128, 139, 144
Particle
  congregations 192
  density 173
  distribution parameters 191

downstream
  coarsening 188-189, 196
  fining 188-189, 196
  elimination 179, 186
  gradient 190
  movement 173
  shape 173, 179
  size 2, 109, 116, 124, 173, 186, 188-195
  sorting 116, 179, 186, 191
  transport
    relationships 186
    velocity 186, 188
Pasture 136-137
Patterned ground 9
Peat 59, 114, 142-143, 149, 151, 153, 156-157, 159, 262
  Blanket 11, 114, 151, 157
  bogs 63, 150, 154, and see Bog
  drainage 66, 158-159
  erosion 11, 143, 268
  growth 157, 160
  hummocks 154
  -lands 2, 149-151, 155-159, 262, 270
  Primary 149, 157
  'recurrence surface' 157-158
  'retardation layer' 157-158
  Secondary 149
  surface 152, 154, 156
Pebble 62, 262
  Arenaceous 206, 208
  durability 208-211
  fraction 203
  fragmentation 209
  percentage composition 203-204
  sampling 205-206
  shape 203, 220
Peneplain 8
Percolation 59, 82, 87, 137, 158
Periglacial activity 8, 11
Permeability 88, 97, 256
Petrographic changes, Downstream 173-178, 210
pH 17, 35
  Soil 34
Phenocrysts 8
Phreatic zone 59
Picea sitchensis (Sitka spruce) 11, 13, 14
Piezometer 82

Pinnacles, Limestone 33
Pits 12-15, 81-82, 84, 185
Plagioclase 11
Planar lamination 183, 186
Planation surface 8
Plantations 11, 21, 73, 80, 96, 114
Plasticity 115-116, 124
  chart 113, 116
  index 113-114, 117, 123
Plastic limit 114-117, 119, 123
Pleistocene 8, 269
Pluton 6
Point bars 182, 207, 220, 228, 232
  deposits 203, 206
Pool-
  hummock complex 151
  riffle 216, 220, 227, 229
Porcellaneous band 53-54, 60
Porosity
  Biopore 96, 262
  Peat 156
  Rock 32, 53, 133, 210
  Soil 76, 78, 86
Potassium 35, 37-38, 40-41, 44, 62, 67
Potholes 51, 56
Pre-Carboniferous floor ('basement') 53, 55, 61-62, 64-65
Precipitation 9, 24, 28, 33, 35-36, 46, 51, 59, 62-64, 76-78, 87, 94, 96, 101, 109, 111, 113, 120, 123, 128, 132, 142, 144, 155-156, 256
  Effective 127, 246
Primary structures 179, 186, 192-196
Productivity 75
Pteridum aquilinum (bracken) 13-14, 114

Quarrying 121
Quartz 11
Quartzite 94
Quaternary
  deposits 137
  history 221
  ice 129
  period 140

Radiation 75
Rain 5, 101, 168, 193
  -days 120

drop splash 2, 34, 43, 46, 76, 106, 142, 182
fall 9, 26, 33, 35, 44, 97, 101, 120, 137, 142, 149-150, 155, 158, 168, 171, 182, 184, 193, 243, 246, 255, 270
  efficiency 102
  erosivity 101
  events 78, 80, 102, 155
  intensity 9, 34, 44, 46-47, 73-77, 80-82, 84, 88, 94, 168, 182, 184
  kinetic energy 76
  Tropical 74, 77, 83, 86, 88
forest 73, 75-79, 86, 88-89, 262
  floor 2, 88
gauges 14
storm 9, 18, 22, 88, and see Storm
water 2, 35, 39, 41, 63, 274
Rand 151, 158
Randomization 276-277
Rating curves, Sediment 129, 132, 137, 144
Reaction rates 1
Regolith 1, 10-11, 13, 17, 96, 103, 124, 215, 262
Rejuvenation 258
Rendzina 112, 114
Re-precipitation 46
Reservoir 17, 22, 127, 196, 268
  Underground 63-64
Residence time 26, 67
Resurgences 54, 61-65, 166-167, 170
Rhacomitrium lanuginosum 153
Ridge-and-trough features 121-124
Riffle 228, and see Pool-riffle
Risings 52, 54, 56, 58-59, 62, 63-65, 67, 164-165, 167, 170-171
River 22, 28, 165, and see Channel, Discharge, Flow, Stream
  diversions 241, 250-251
  history 241, 250
  Underground 165-166

Rodents 98-99
Root
  mat 97
  penetration 97
  surface 78
Rooting 24, 97
Runoff 1, 23, 31, 36-38, 42-43, 47-48, 73, 76-77, 81, 102, 114, 120, 127, 129, 142, 158, 160, 163, 168, 184, 243, 246, 255-256, 262
  Ephemeral 179, 183-185, 187-188, 193, 195-196

Salts 5, 275
  airborne 134
Sample collection 14-15, 17, 35-36, 81-82, 101, 116, 129, 153, 185-186, 206
Sand 2, 61, 113-114, 116, 123, 182, 215, 220, 265
Sandstones 51-53, 94, 96-97, 110, 114, 133-134, 140, 164, 182, 204, 208, 251-253, 255
Scale 2-3, 122, 124, 127-128, 140, 144-145, 168-169, 173, 220-221, 223, 242, 257, 263, 270
Scars 52-54, 62
  Valley-side 220, 231
Scree 120-121
  Cemented 168-169
Scour
  -fill processes 179, 189-190, 196
  surface 184, 220-221, 228, 230
Sea-level 242
Seasonal variation 21-22, 26, 28, 33, 42-46, 80-82, 97-98, 100-106, 117, 119, 144, 150, 154-155, 163-164, 182, 190, 265, 274, 278
Sediment 2, 6, 28, 101, 270
  availability 133
  budgets 127, 133, 136
  concentrations 102-103, 106, 129, 132-134, 140-141, 144
  discharge 102-106
  dispersal 179
  disposal 179
  entrainment 182, 188, 192

Lagoonal 191-192
load 128-129
Offshore 189-190
particle population 173
Shoreline 191
skewness 189, 191
sorting 179, 184, 192-196, 220
 source 97-98, 109, 136-137
 suspended 28, 127-128, 132, 136-137, 215
  loss 129, 131-132, 136-137, 139, 268
  pulse 102
  yield 94, 127-128, 135, 140, 144
 translocation 99-102
Sedimentary
 environment 186, 194
 parameters 173
Sedimentology 173
Seepages 13, 15, 97, 114, 167
 Influent 184
Shakeholes 59
Shale 51, 53, 56-57, 59, 93, 103, 164, 204, 251, 253
Shear
 boundary 188
 force 173, 188
 Rock 55
Shore
 line 191, 194
 processes 191
Shrubs 75
Silica 24, 76, 262
Silicate minerals 26, 28
Silicic acid 8
Silicon 6, 18-28
Silt 96, 113, 123, 181-182, 221, 228
Sinkholes 51-53, 59-60, 62-63
 Peatland 158-159
Slates 220
Slides 109
Slope
 angle 9, 26-27, 93, 101, 109
 base 21
 Break of 9, 24, 158-159
 deposits 103
  Bedded 11
 pantometer 121
 profiles 121-122, 272
 Scarp 94
 steepness 112
 transect 10-13, 18
 wash 78

Slopes 6, 9-10, 13, 33, 37, 76-77, 79, 81, 88, 93-99, 103, 109-110, 114, 116, 120-123, 132-133, 137, 140, 142, 150, 159, 169, 182, 188
Slumps, Rotational 120
Snails 40-41, 43, 47, 262
Snow 111, 150-151, 275
 bank 38
 fall 9, 142
 melt 67, 171
Sodium 35, 62, 67, 134
 chloride 134
Soil 11, 34, 76-77, 79, 96, 112-116, 123, 163, 265
 adhesion 114
 aeration 75
 aggregates 96-98
 Bare 98-99, 101-102, 106
 catena 112, 272
 cohesion 114-115
 compaction 96
 consistency 114
 cover 31, 35, 38-39, 48, 59, 67
 creep 109, 114, 116, 124
 crust formation 96
 drainage 94, and see Infiltration, Interflow
 erodibility 97, 99
 erosion 93, 127
 fall 97-98
 fauna 96
 field moisture content 114-116, 120, 124
 flow 115, 124
 heterogeneity 277
 horizons 13-15, 18, 76, 82, 96-98, 105, 114
 humus 93
 loss 93
 mechanics 123-124
 moisture 18, 59-60, 75, 87, 97, 109, 114-115, 144, 229
 tension 84, 115
 Organic 39
 particles 2, 76, 109
 pillars 182
 pit 15-17, 20, 265
 podzolic 11, 114
 profile 13, 26, 81, 96, 113
  discontinuities 17
  Exposed 97
  Truncated 96
 samples 116
 saturation 82, 88, 97
 science 3

shrinkage 97
Skeletal 140
stability 109
structure 15
surface 77, 86, 88, 96-97
swelling 97
texture 17, 26, 76, 96-97
  113-116
  Sandy 76
  Stony 96
types 11, 96, 112
water 1, 5-6, 14, 17-18,
  21, 270
Solar energy receipt 33
Solifluction 8-9, 59
Solutes 5, 14, 31, 36-38,
  41-46, 64-66, 68, 133, 163,
  270
  budget 133
Solution 32-33, 41, 46-47,
  62-63
Solutional processes 2, 8,
  26, 33-34, 43, 51
  Underground 57
Sorting processes 179, and
  see Sediment sorting
Sparite 53
Spatial
  analysis 271-272
  distribution 94, 179,
    271-272
  pattern 189
  sequence 271-272
  variability 3, 21, 31, 36,
    137, 211, 215
Specific conductance 17, 35-
  38, 40, 44-46
Sphagnum 153-154, 156-157
  papillosum 159
Springs 2, 13, 15-16, 22-23,
  26-28, 97, 110, 167, 171,
  269-270, 274, 277
Statistical methods 276-277
Stemflow 5, 14, 18, 20-21,
  24, 76, 100
Stones 99
Storage capacity 160, 185
Storm 33-34, 36, 38, 42,
  44, 47-48, 64, 76, 81-
  83, 87, 102, 132-133,
  141, 182, 184
  concentrations (of sediment)
    128, 137-138, 140-141
  drainage 127
  intensity 42, 44, 47, 87
  magnitude 42, 47, 87
  movement 83-86, 184, 193
Stratigraphy 180, 196

Stream 53
  channel 17, 78, 93-94,
    and see Channel
  bed 105, 173, 186, 192
    Sand- 173, 187
  density 137
  deposits 179
  flow 5, 88, 149, 164,
    184, 215, 265, and
    see Flow
  gauging 17, 86, 129,
    184, 246
  head 78-79
  incision 181-182
  length 243
  'relief' 243
Strike 53
  -etching 182, 196
Stromatolites 182
Subsoil 5, 96-97
Subsurface flow 5, 78, 82,
  86, 136, 184, and see
  Flow, Interflow
Sulphuric acid 66
Sunlight 35, 151
Surface
  flow 2, 13, 163, 165,
    168-169
  drainage 51-52, 59
  lowering 47-48
  runoff 5, 34, 51, 73,
    78, 86, 136, 140, 166,
    168, 171
  wash 182
  water 3, 247
Suspended load 127, 192

Temperature
  Air 9, 15, 33, 39, 76-
    77, 111, 127-128,
    150, 163-164, 169-
    170, 270, 274-275
  Rock 35
  Water 15-17, 22-23, 27,
    39, 65-67, 163-167,
    169, 274
Tensiometer 82, 84
Thalweg 223, 230
  step-length 221
Theory 1, 3, 41, 272, 275
Throughfall 5, 14, 18,
  20-21, 24, 35, 88
Throughflow 97, and see
  Interflow
Time 3, 269-271
  lag 84-85, 88, 97, 127,
    171
  Post-glacial 11, 150,
    156, 247, 257

sequence 121-122, 270
Tin streaming 9, 11
Tipping buckets 81, 84, 87
Toppling failure 120
Torque 192
Tors 1, 9
Tourmaline 8
Tower karst 31-33
Traction
  carpet 192
  threshold 188
Transmission losses 185
Transpiration 5
Transportation 1, 28, 127, 143, 173, 179, 209, 211, 215, and see Sediment, Soil
Slope 99-102, 127, 189
Travel 272-275
Tree 35, 75, 77
  Beech 100
  canopy 35, 93
  fall 76
  roots 77-78, 96
  spacing 14
  trunks 76, 100
Tributary 174, 184-185, 196, 203, 206, 208
  confluence 185, 192-193, 195, 204, 242
  sediment size 189
Troughs 81-82, 84, 274
Tuff 182
Twigs 5
Turf 14, 20
  runoff 16, 24

Unconformity 54, 61, 67
Undercutting 33, 97
Underground drainage systems 51, 54, 59, and see Cave

Valley
  bends 168, 220, 227
  fill 215, 228
  floor 9-11, 13, 17, 52, 110, 114, 151, 182, 215, 220
  side-slope 169
    bluffs 216, 228, 230
    constraint 182, 188, 215
    scars 220
    walls 215, 237
  Upland 2, 235
Variable source area 73
Vegetation 10-11, 14, 18, 24, 75-77, 96-97, 106, 109, 114, 120, 124, 128, 131-133, 142, 149, 151, 221, 262, 268

canopy 5, 13
changes
  Post-glacial 11
  Man-induced 11
Vermiculite 11
Vitex siamica 35

Water
  chemistry 5
  falls 184, 253, 273
  Infiltrating 6
  logging 97
  levels 152-155, 206, 262
  table 60, 149, 153-156, 158, 168
Wave
  action 179
  generation 191
Weathered debris 127, 263
Weathering 6, 96, 252
  Chemical 1, 28, 31, 127, 261-262
  Tertiary 8
  Tropical 1, 79
Wells 82, 153, 274
Wetting front 18
Wind 33, 78, 80, 94, 98, 101, 150, 191, 265
Windypits 110, 121-122
Winnowing 188, 209
Woodland 11, 94
  Beech 93, 96, 103
  Coniferous 11
  Oak/beech 93, 96
  Spruce 96
  Pine 96
  Carr 157
Worms 78, 98-99
  casts 99

Yoredale strata 51-52, 56, 63, 171, 204-206